Make:

Volume 11

technology on your time ™

Alt.Vehicles

48: Urban Guerrilla Movie House
Your own DIY drive-in. By Mister Jalopy

58: The Little Cart That Couldn't
Is a wind-powered vehicle claim just a bunch of hot air?
By Charles Platt

64: Swiveling Balcony Hoist
Take the sting out of walk-up apartments. By Matthew Russell

67: Rolling Solar
From junker to sun-charged car. By Ben Shedd

68: Granny's Nightmare Chopper Bike
Chop an old ladies' bicycle into something evil.
By Brad Graham

73: Swing & Wrong-Way Bikes
Trick cycles from Cyclecide. By Paul Spinrad

74: U-G-L-Y Your Bike
To deter thieves, camouflage your bicycle as a piece of crap
while keeping it a first-class ride. By Rick Polito

76: Rock the Bike
Social biking with Fossil Fool and the Juice Pedaler.
By Paul Spinrad

80: iPod Bike Charger
A sidewall dynamo powers both lights and tuneage.
By Mark Hoekstra

82: Stokemonkey Makes It Easier
An electric motor linked to existing bicycle gears turns any
bike into a sell-your-car-already vehicle. By Rick Polito

84: The Year People Learned to Fly
Celebrating the 30th anniversary of the flight of
the *Gossamer Condor*. By Ben Shedd

PEDALING THE FRIENDLY SKIES
In 1977 Paul MacCready's *Gossamer Condor* took flight with a 120-pound cyclist as both the pilot and the engine, marking the first human-powered flight in history.

84.

ON THE COVER: Photographer Robyn Twomey set up on MAKE Editor-in-Chief Mark Frauenfelder's lawn to shoot Mister Jalopy's Urban Guerrilla Movie House. Tooth-decaying snacks were used to bribe the kids to stare at a blank screen for extended periods of time.

Columns

10: Welcome
By Gareth Branwyn

15: Make Free
Nearly ten years after the invention of Napster there's more
file sharing than ever. By Cory Doctorow

24: Hands On
Now that "digital carpentry" has come to exist, how do you
make it authentic? By Bruce Sterling

38: Heirloom Technology
The Chinese make fuel from coal dust, farm waste, scrap
wood, and a bit of the local dirt. By Tim Anderson

44: Making Trouble
Everything I know I learned from two wheels and a frame.
By Saul Griffith

92: Personal Fab
A CNC milling machine for less than a grand. By Tom Owad

182: Retrospect
The 1977 Gulf of Alaska Baidarka Expedition.
By George Dyson

Vol. 11, August 2007. MAKE (ISSN 1556-2336) is published quarterly by O'Reilly Media, Inc. in the months of March, May, August, and November. O'Reilly Media is located at 1005 Gravenstein Hwy. North, Sebastopol, CA 95472, (707) 827-7000. SUBSCRIP-TIONS: Send all subscription requests to MAKE, P.O. Box 17046, North Hollywood, CA 91615-9588 or subscribe online at makezine.com/offer or via phone at (866) 289-8847 (U.S. and Canada); all other countries call (818) 487-2037. Subscriptions are available for $34.95 for 1 year (4 quarterly issues) in the United States; in Canada: $39.95 USD; all other countries: $49.95 USD. Periodicals Postage Paid at Sebastopol, CA, and at additional mailing offices. POSTMASTER: Send address changes to MAKE, P.O. Box 17046, North Hollywood, CA 91615-9588. Canada Post Publications Mail Agreement Number 41129568. CANADA POSTMASTER: Send address changes to: O'Reilly Media, PO Box 456, Niagara Falls, ON L2E 6V2

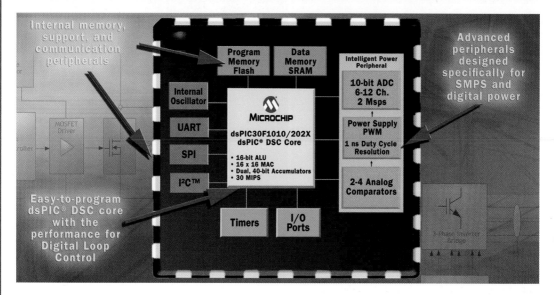

Make: Projects

Retro R/C Racer

Use scrap sheet metal and pop rivets to build a model 1930s British Midget racer with modern R/C capabilities. By Frank E. Yost

94

Vacuum Former

Mold light, durable, and cool-looking 3D parts in your kitchen. By Bob Knetzger

106

Rotating Bird Feeder

Get awesome avian photos by moving your camera closer to the birds — and rotating them for the perfect pose. By Larry Cotton

116

Solid Geometry

Learn about the five Platonic solids, and then build a stellated dodecahedron lamp. By Charles Platt

164

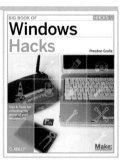

Make:
Volume 11

technology on your time ™

Maker

16: Made on Earth
Snapshots from the world of backyard technology.

26: Hammer Time
Making the antiques of the future at Black Dog Forge.
By Kirsten Anderson

30: Big Blowhards
Who will be the first to make a machine that propels a pumpkin more than a mile? By William Gurstelle

34: The Maker State
Safe working practices give you the freedom to attempt projects on the edge. By William Gurstelle

36: Drawbot Love
How a bunch of us made our own drawing robot. By Bre Pettis

40: Proto: Power Tripping
High-voltage engineer Greg Leyh builds the largest Tesla coils in the world. By David Pescovitz

46: 1+2+3: Ten-Second Stomp Rocket
Make a juice-bag-powered rocket in less time than it takes to drink it. By Emma Wagstaff

86: Spirits Guy
How Lance Winters went from basement moonshiner to celebrity vodka distiller. By Benjamin Tice Smith

126: 1+2+3: Simple Motor
Make a spinning motor with a minimum of parts.
By Cy Tymony

172: Blast from the Past
Orange crate racer. By Mister Jalopy

DIY 127

127: Circuits
Breadboard rack, big kid night light, net data meter.

138: Science
MAKE Controller weather fetcher.

141: Home
Low-cost spherical speaker array, free VoIP, plush irradiated sirloin.

151: Workshop
Learn which battery to use for your project.

154: Outdoors
Put electronic crickets and fireflies in outdoor solar lighting.

157: Citizen Scientist
Track your ticker with a homemade electrocardiogram machine. By Dr. Shawn

160: Howtoons: Wheel of Life
By Saul Griffith, Joost Bonsen, and Nick Dragotta

162: MakeShift: Hot Electrical Problem
By Lee D. Zlotoff

169: Toys, Tricks, and Teasers
The oscillating beam machine. By Donald E. Simanek

174: Toolbox
The best tools, software, publications, and websites.

181: Aha! Puzzle This
Garden gnomes and wacky races. By Michael H. Pryor

185: Tips and Tricks
Scrambled eggs from an espresso maker and more, plus a new visual how-to.

190: Maker's Calendar
Our favorite events from around the world.
By William Gurstelle

192: Homebrew
The train-schedule alarm clock. By Greg McCarroll

STILL LIFE
Inside Lance Winters' high-grade hooch factory at Hangar One in Alameda, Calif.

86

Make:
technology on your time™

EDITOR AND PUBLISHER
Dale Dougherty
dale@oreilly.com

EDITOR-IN-CHIEF
Mark Frauenfelder
markf@oreilly.com

MANAGING EDITOR
Shawn Connally
shawn@oreilly.com

ASSOCIATE MANAGING EDITOR
Goli Mohammadi

SENIOR EDITOR
Phillip Torrone
pt@makezine.com

PROJECTS EDITOR
Paul Spinrad
pspinrad@makezine.com

STAFF EDITOR
Arwen O'Reilly

COPY CHIEF
Keith Hammond

EDITOR AT LARGE
David Pescovitz

VIDEO PRODUCER
Bre Pettis

CREATIVE DIRECTOR
Daniel Carter
dcarter@oreilly.com

DESIGNER
Katie Wilson

PRODUCTION DESIGNER
Gerry Arrington

PHOTO EDITOR
Sam Murphy
smurphy@oreilly.com

ONLINE MANAGER
Terrie Miller

ASSOCIATE PUBLISHER
Dan Woods
dan@oreilly.com

CIRCULATION DIRECTOR
Heather Harmon

ACCOUNT MANAGER
Katie Dougherty

MARKETING & EVENTS COORDINATOR
Rob Bullington

MAKE TECHNICAL ADVISORY BOARD
Evil Mad Scientist Laboratories, Limor Fried, Joe Grand, Saul Griffith, Bunnie Huang, Tom Igoe, Steve Lodefink, Erica Sadun

PUBLISHED BY O'REILLY MEDIA, INC.
Tim O'Reilly, CEO
Laura Baldwin, COO

Visit us online at makezine.com
Comments may be sent to editor@makezine.com

For advertising inquiries, contact:
Katie Dougherty, 707-827-7272, katie@oreilly.com

For sponsorship inquiries, contact:
Scott Fein, 707-827-7102, scottf@oreilly.com

For event inquiries, contact:
Sherry Huss, 707-827-7074, sherry@oreilly.com

Contributing Editors: Gareth Branwyn, William Gurstelle, Mister Jalopy, Brian Jepson, Charles Platt

Contributing Artists: Caren Alpert, Scott Beale, Roy Doty, Nick Dragotta, Dustin Amery Hostetler, Timmy Kucynda, Terry Leonard, Tim Lillis, Tom Parker, Charles Platt, Nik Schulz, Damien Scogin, Jen Siska, Jonathan Sprague, Robyn Twomey

Contributing Writers: Kirsten Anderson, Tim Anderson, Joost Bonsen, Annie Buckley, Dr. Shawn Carlson, Larry Cotton, Cory Doctorow, Andrea Dunlap, George Dyson, Lenore Edman, Limor Fried, Brad Graham, Saul Griffith, Mark Hoekstra, Tom Igoe, Walter Kalwies, Bob Knetzger, Dave Mathews, Greg McCarroll, Steve Nalepa, Meara O'Reilly, Ross Orr, Tom Owad, John Edgar Park, Rick Polito, Michael H. Pryor, Douglas Repetto, Matthew Russell, Ben Shedd, Donald E. Simanek, Benjamin Tice Smith, Darel Snodgrass, Bruce Sterling, Rebecca Stern, Cy Tymony, Emma Wagstaff, Daniel Weiss, Megan Mansell Williams, Frank E. Yost, Michael F. Zbyszynski, Tom Zimmerman, Lee D. Zlotoff

Interns: Matthew Dalton (engr.), Adrienne Foreman (web), Arseny Lebedev (web), Jake McKenzie (engr.)

Customer Service cs@readerservices.makezine.com
Manage your account online, including change of address at:
makezine.com/account
866-289-8847 toll-free in U.S. and Canada
818-487-2037, 5a.m.–5p.m., PST

Contributors

Bob Knetzger (*Vacuum Forming project*) has been making fun stuff all his life: first as a kid with his trusty Vac-U-Form, then fresh from design school on staff at Mattel Toys, and for the last 25 years, as Neotoy with his business partner, Rick Gurolnick. Bob lives near Seattle with his wife and two kids, and plays steel guitar with The Swains (swainscountry.com) when he's not molding plastic. His heroes include creative Californians Ed "Big Daddy" Roth, Leo Fender, Charles and Ray Eames, and Sneaky Pete Kleinow.

Limor Fried (*DIY Batteries*) likes to design nifty electronics and release them as open source projects, such as a monophonic synthesizer and a PLL-controlled cellphone jammer. She has somehow managed to turn this hobby into a viable business by selling kits. Her biggest pet peeve is connecting power to ground. ladyada.net

Kirsten Anderson (*Hammer Time*) is "a cross between Morticia Addams and that weird old neighbor lady who always wants to know everyone's business." Current interests: art of any stripe, wildlife documentaries, mutant anomalies, cryptozoology, elephant culture, funeral practices throughout history, deep sea creatures, lawn bowling, donut eating, "sustainable living," shopping for arcane trinkets on eBay, and horror magazines. She owns and runs Roq la Rue Gallery in Seattle, where she lives with her "awesome husband and two weirdo Siamese cats." She "could eat pad Thai morning, noon, and night," and often does.

Arseny Lebedev (MAKE web intern) grew up in Moscow, Russia, and now lives on the East Coast, working at MAKE's New York office and studying business and technology at a local university. "I bring my own grocery bags to the store, and when I was younger I would make UFOs out of cotton balls, wire, alcohol, and dry-cleaner bags. I'm a big geek, and I love to mess with computers, video games, and food."

Rebecca Stern (*DIY Plush Sirloin Night Light*) describes herself as "a new media artist working at the intersections of diverse fields to make culturally significant projects." That doesn't distract her from mechanically automating her analog thermostat with her roommate to adjust the temperature using a microcontroller, a motor, and a rubber band. She now lives in Tempe, Ariz., where she's a Ph.D. student in ASU's Arts, Media and Engineering program. She loves "Michael Pollan's books, cooking, knitting, traveling, riding my bike, camping, tinkering with computers, and (of course) making things."

Caren Alpert (*Maker profile* photography) loves to cook, travel, and exercise so she can eat chocolate and mac-and-cheese whenever possible. A photographer whose work has been featured in numerous magazines and cookbooks, she's currently working on a monograph about architect Morris Lapidus. She lives in San Francisco with her pet rabbit, Moose, and her favorite color is eggplant.

When **Benjamin Tice Smith** (*Maker profile*) isn't working as a photo editor, he's usually making things out of wood, metal, and concrete, or rebuilding and keeping old audio equipment, cars, and motorcycles running. He brewed a lot of beer and mead at one time, and loves a good cocktail, but his favorite tipple is Scotch. He lives in Oakland, Calif., with his wife, two children, and two dogs. Now that his children have discovered *Star Wars* and have assigned the family characters for Halloween, he (Darth) is trying to figure out costume designs for everyone.

Welcome

MAKER FAIRE IS TOO MUCH

By Gareth Branwyn

Editor's Note: MAKE Technical Advisory Board member Gareth Branwyn and his son, Blake, ran the Mousey the Junkbot Workshop in the MAKE area at Maker Faire. The following excerpt from Gareth's blog, streettech.com, captures his experience at the Bay Area Maker Faire in May 2007.

EVERYTHING AT MAKER FAIRE IS cranked to 11: the size of the event, the creativity of what's being presented, the excitement of the fairgoers, the diversity of the people who show up. So, *you* end up on 11. I heard that this jacked amperage was experienced by both fairgoers and presenters alike. The common chant went something like: "This is so awesome. I love it! There's too much! I'll never get to see it all."

As workshop presenters, Blake and I saw little of the Faire. The first day, we did open-ended workshops, selling Mousey parts bundles and then helping people build them at workstations we'd set up. That was probably the most tired I've ever been in my life. The second day, we ran three one-hour workshops. That gave us some time to wander around and see some of the Faire.

Our Mousey workshops went very well. We created two parts bundles (put together by the fine folks at Solarbotics [and still available at store.makezine.com]). We made a quicker, easier "car kit," but ended up only selling three of them. Everyone bought the full Mousey, and a surprising number of people actually sat down and started the build right there in the MAKE area. People were at the workstations for several hours. My favorite was a woman who saw the mousebots, really liked them, and said: "You know what? This is really out of my comfort zone, but I'm going to do it anyway. I think I need to challenge myself more." And she bought a parts bundle, chose an old mouse, sat down, and dove right in. There were a lot of kids with their parents, moms and dads alike, working together, which was nice to see.

Other highlights of the show for me were Mister Jalopy's talk and his Urban Guerrilla Movie House on wheels (*to build your own, see page 48*). One of my new favorite words is *hilaritas*, which means "profoundly good-natured, full of mirth." It's more than being friendly, more than being funny. Mister Jalopy is full of hilaritas.

One of the things that really struck me about the Faire was the impressive diversity of the attendees. The MAKE ethos really does appeal to an extremely broad range of people. The staggering diversity and creativity on display were also evident in the vehicles that freely circulated throughout the fairgrounds.

It was the most insane, and insanely great, fleet of conveyances I've ever seen: all manner of odd vehicles, from electric bikes and cars, to pedaled recumbents, homemade Segways, a solar-powered motorcycle, and a guy riding a motorized unicycle while holding a regular unicycle in front of him. And then there was the chariot pulled by a Roman centurion robot, and a covered wagon pulled by two robotic horses. It was all so surreal, so ethereal, so like a dream, if your dreams were scripted by Salvador Dali and Rube Goldberg.

At dinner on the first night, Mark Frauenfelder and I were talking about the continued, cancerous growth of the American monoculture, as it spreads across the planet like the chocolate pudding blob from a 1950s sci-fi horror flick. There are few regional differences anymore, little local color. The cyberneticist Gregory Bateson is famous for saying "information is difference," and "information is difference that makes a difference." That's what's so scary about our planet-invading monoculture. No difference? No information.

The beauty of the Maker Faire is that it's about crazy, almost fractal, levels of difference. So many people came up to me, looked at my project, which turns a useless analog mouse into a light-seeking robot, and were giddy, almost drunk, with excitement, over all of the monocultural boxes that they saw transcended at booth after booth. "There are so many innovative ideas here!" they enthused. "I can't get over all of the ingenuity, the creativity!" "How did you ever even think up such a thing?" And on and on.

So, in conclusion: run for your lives! And run to the next Maker Faire, if you can.

The next Maker Faire is Oct. 20–21, 2007 in Austin, Texas, at the Travis County Exposition Center and Fairgrounds.

OH SNAP!

A TABLE MADE FROM JUGS.

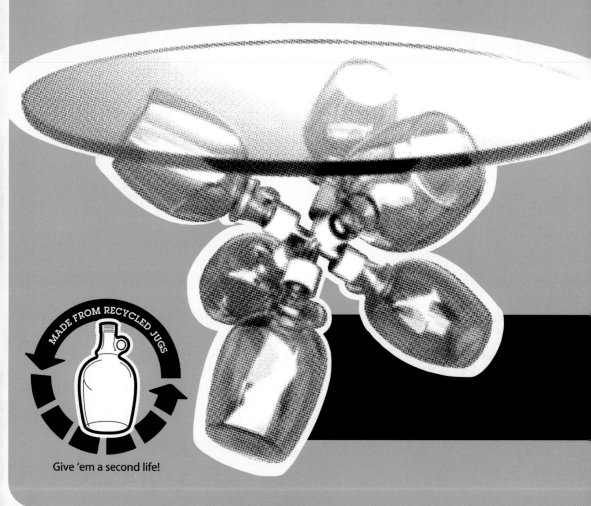

MADE FROM RECYCLED JUGS

Give 'em a second life!

GOOD THINGS COME TO THOSE WHO CREATE.

WOW!

THE BEST PART WAS GETTING ALL MY FRIENDS TOGETHER TO HELP ME EMPTY ALL THE JUGS. GENIUS!

COSMIC!

MMMMM, THAT WINE SURE IS TASTY. AND SO ARE ALL THE THINGS YOU CAN MAKE WITH AN EMPTY JUG. NOW ALL MY STUFF IS JUG FURNITURE!

ALWAYS REMEMBER, from a full jug of Carlo Rossi flows great wine. And from an empty jug flows great ideas. Now go make something!

BOOYAH!

JUGTASTIC!

LOOK!

THIS JUG CLUSTER TABLE WAS DESIGNED AND BUILT BY THE CHARLES FURNITURE TEAM. STRAIGHT OUTTA' DUBLIN! THEY MADE IT WITH SIX EMPTY 4.0 L JUGS, AN ALUMINUM FRAME AND A BUNCH OF INGENUITY. NO LUCK OF THE IRISH NEEDED.

Make Free

ATTACK, DEFEND

By Cory Doctorow

WHY DO WE GET THE FUTURE SO wrong? From *The Jetsons* to *Future Shock*, from Asimov to H.G. Wells, our species appears to roundly suck at predicting the future. Science fiction tells you a lot about the biases of any given writer's era, but precious little about the future we're heading to. (Hence all those wonderful articles from this magazine's 1930s forebears, like *Modern Mechanix*, about the coming world of metal men who will wait on us hand and foot.)

Today, futurists talk about whether our next society will be more global, more automated, more religious, more democratic. They point to the signs in the same way the Mesopotamian extispicists did, claiming to see the future in animal organs.

I think they all get the same thing wrong: they presume that technological change will create a progressive series of epochs, like the scenes in Walt Disney's Carousel of Progress, in which we are taken through four robotic dramas tracing life from the dawn of the electric age to the Marconi era to the Fabulous Forties, and into the "present day" (an embarrassingly awesome vision of the American living room circa 1993 or so).

For this to happen, technology will have to produce more than change — it will also have to produce stability. And that's the most unlikely prediction of all.

From a security perspective, technology usually gives an inherent advantage to attackers. Take earthwork fortifications: defenders needed to put up a perfect bulwark to keep the barbarians out. Barbarians needed to find one weak spot and crash through. Defenders need to be perfect, attackers need to find a single flaw.

That's why it's been so easy to kick the living crap out of the entertainment industry in the cat-and-mouse game of file sharing. Here we are, nearly ten years after the invention of Napster, and there's more file sharing than ever. At the time of this writing, the internet is still chortling up its collective sleeve about the efforts of the Advanced Access Contact System Licensing Administrator (AACS LA) to suppress a 128-bit number that can be used to decrypt some HD DVDs. There are presently more than 2 *million* web pages that contain this number (there were about 100 when the AACS LA sent out its legal threats).

This attack/defend disparity is why it was so easy to get the AACS keys out of a DVD player in the first place. To keep anti-copying keys secret, they have to be perfectly protected in every single device manufactured and distributed. To extract the keys, one need only discover a single vendor that made a single mistake in its production.

What's sauce for the goose is sauce for the gander. Technology lets the Recording Industry Association of America automate its lawsuit process and attack 700-plus Americans every month. Technology lets al-Qaeda form a loose, undefined network that can wreak terrible havoc. Technology lets hackers hijack computers and turn them into "botnets" of spam-sending, denial-of-service-launching zombies.

The attack-defend disparity is great news if you don't like the status quo, but it means that any victory is bound to be short-lived. No sooner do you dismantle your enemy's fortress and put up your own than someone comes along and does to you what you just got through doing to him.

And that's what's wrong with today's futurism. The industrial revolution wasn't just a revolution: it was a transition. The world moved from agrarianism to industrialism. But the information revolution isn't a transition; it's a continuous, permanent revolution. The one thing the future won't have is enough of a status quo to matter. Constant change will be the hallmark of human history.

Not that I'm complaining, mind you. When your attention span is as short as mine, constant change is totally addictive. But the next time some wag talks about a 50-year biotech plan, or the next 100 years of nanotech, ask yourself: if these technologies are so darned disruptive, won't they disrupt themselves?

Cory Doctorow (craphound.com) is a science fiction novelist, blogger, and technology activist. He is co-editor of the popular weblog Boing Boing (boingboing.net), and a contributor to *Wired*, *Popular Science*, and *The New York Times*.

Global Bus

What kind of a Volkswagen bus rolls perfectly well but doesn't drive? One molded into a perfect sphere by sculptor **Lars Fisk**, a former art director for the jam band Phish.

Fisk first conceived of orb art in the 1990s while driving through the Green Mountains on Vermont's I-89 in a state of self-described "road hypnosis." Staring at the passing pavement, he imaged the tarmac rolling up beneath him into a spherical form that could be placed somewhere — like on the floor of a gallery. So began a decade-long obsession with balls.

"It's an unlikely shape for any useful object to take," Fisk says. "They're unusual for terrestrial use because they just end up rolling away."

Fisk set to work on the concrete and asphalt *Street Ball*, complete with painted traffic lines. He went on to make a sod-covered *Field Ball*, as well as a *Tree Ball* made from a maple log. *Barn Ball* incorporated hay, and *Sink Ball* ... well, you get it. His streak of spheres inspired by motorized objects included *John Deere Ball* and *UPS Ball*, which had such an authentic brown-and-gold paint job that it

was mistaken for company property.

Fisk based his most ambitious ball of all on a 1967 Volkswagen Samba Deluxe. To build the globular bus, he scavenged parts — seat, steering wheel, emblems — from junkyards and anywhere else he could find them. The dashboard ashtray came from Australia; some Beetle parts snuck in. Fisk freely describes how he formed the windshield through a process called slumping, which involves delicately melting a flat pane into a mold so that it relaxes into place. But exactly how he shaped the metal shell is a trade secret.

"I like the balls to be kind of elusive," he says. "The mystery is part of the interest. People mistake these things for actual objects. They cannot place them in their world. They're not sure if the balls are supposed to be taken as sculpture or what."

—*Megan Mansell Williams*

>> **Fisk's Work at Taxter & Spengemann Gallery:**
taxterandspengemann.com

>> **Fisk's Artist in Residency Program:**
treyanastasio.com/nonprofit

Super Stick

Jason Torchinsky has fond memories of gazing intently at a television screen animated by red and yellow explosions and stiff-jointed figures, two hands at the ready on his Atari joystick.

So when John Gibson, co-curator of the I Am 8-Bit art exhibition inspired by 80s video games, called with a challenge — to make a large, interactive piece for the show in just over a month — Torchinsky couldn't refuse.

He chose the much-loved black, yellow, and red Atari 2600 controller as inspiration, and the resulting piece, a fully functioning, larger-than-life joystick, was exhibited at Gallery 1988 in Los Angeles this spring. Roughly 1,500 people attended the opening, climbing on the giant toy to play a game projected on the patio.

Torchinsky built the 5-foot stick in his driveway under a tarp and, despite one nerve-wracking rainy night, the setup was successful. After measuring each part of an original Atari (though not the one he played as a kid; he's "pretty sure Mom threw it out a while back"), Torchinsky drew diagrams

of the structure and electronics, multiplying each measurement by 15. He took the drawings to several cabinetmakers before finding one, Dan Phill, willing to take on the job.

The pieces were measured and cut from fiberboard in a two-part construction: the bottom piece holds electronics, the top holds a large joystick surrounded by a coiled hose and the frightfully cheerful red "fire" button. Five switches make the video action happen: up, down, right, left, and fire. Each is attached to a metal plate that connects to the appropriate pin on the controller as someone moves the joystick from above.

"I like things that people can engage with and have fun," Torchinsky says, casually buffing scuff marks on the meticulous paint job with his hand. Though game-players' fun took a toll on the surface, the whimsy and humor of the piece survived intact.

—*Annie Buckley*

≫ **Giant Joystick:** jasontorchinsky.com

Photograph by Annie Buckley

Photography by Simon Jansen

Fire-Cooled Brew

New Zealander **Simon Jansen** has all the bona fides of an alpha maker. A software engineer and classic car restorer, he's got a half-built R2-D2 and a custom minibike he made from scratch. He achieved geek fame with his ASCII animation of Star Wars scenes (asciimation.co.nz), which practically defined obsessive attention to detail.

But a jet-powered beer cooler? This bloke operates on a whole 'nother level of absurdity.

Jansen set out to make the holy grail of many a maker: the homemade jet engine. In his Auckland garage, he welded his own combustor, bolted it to an old turbocharger, and added a leaf blower for air flow and a propane tank (sans regulator) for fuel.

The trickiest part was the oil system, which must maintain critical lubrication pressure: "I used an oil pump from an old Ford Escort Mark 1, driven by the motor and gearbox from a cheap 12-volt rechargeable drill!"

Don't try it at home without an exhaust temperature gauge that goes to 1,000°F and an rpm meter that hits 100,000. But bloody hell! It worked, with the head-splitting roar that jet hobbyists live for. "Incredibly loud," Jansen recalls fondly. "You can hear the air being ripped apart as it is sucked into the turbine. I was grinning for days."

From adversity came the real breakthrough. Jansen's jet burned propane so fast that the tank rapidly iced up, dropping the fuel pressure. So he stood the tank in a tub of warm water. When a colleague remarked that the iced water could then chill beverages — eureka!

Jansen says beer and dangerous machines don't mix, so he abstains from the frosty bevvies until he's finished playing with the engine. Ever the tinkerer, he has stripped down and rebuilt the jet beer cooler several times. "The latest iteration should be more self-contained and portable," he promises. "I've been telling the mates at the office we'll fire it up in the car park." —*Keith Hammond*

≫ **Jet-Powered Beer Cooler:** asciimation.co.nz/beer

➕ **More Homemade Jets:** junkyardjet.com

The Old-Fashioned Future

If you love the look of movies such as *The Nightmare Before Christmas* and *The City of Lost Children*, there's a sculptor out there after your own heart. In fact, none of **Stéphane Halleux**'s darkly whimsical renderings — robotic wheelchairs, squat submarines, armored cars, men with mechanical bat wings, animal soldiers wearing hand-stitched leather gas masks, a 30-inch cyclist whose suit includes electrodes that generate energy — would be out of place in these fantasy flicks.

Two years ago, the former comic-book illustrator got sick of drawing "uninteresting things for other people" and started building the lanky-limbed, roboticized characters he enjoyed doodling — ominously cartoonish forms that are futuristic yet also recall the past.

Now, Halleux works on his sculptures full time out of his home in the Belgian countryside, where he lives with his partner and their two boys. He sketches the rounded and slumped shapes, then coaxes the creatures from wood, metal, and hide, using scrounged antique parts for appendages

and each of the fine details.

"It's as if these old, saved elements, full of history, were giving a soul to the final work," Halleux says of his recycled materials. "I like crazy mixtures, unlikely associations, advanced technology mixed with mechanisms of long ago."

The seemingly mechanical conformations give the impression that they are capable of rolling, taking off, or — in the case of the cyclist — pedaling. But actual movement is all in the spectator's mind, Halleux says.

"Each one has an invented history. If I really wanted to animate them seriously, the form and atmosphere would suffer. I think imagination is stronger than a working light or a turning propeller."

—*Megan Mansell Williams*

≫ **Halleux's Collection:** www.stephanehalleux.com

Photograph by Muriel Thies

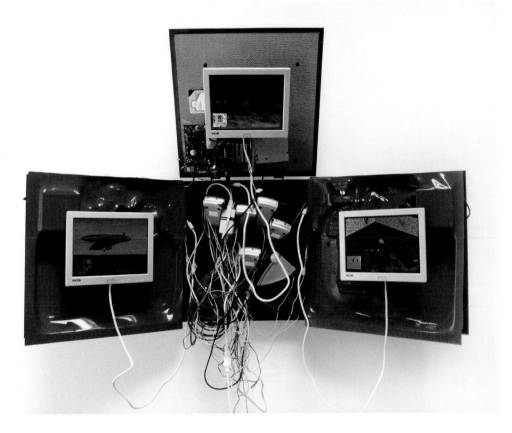

Games That Games Play

Video games lead lonely lives. The characters are trapped inside infinitely repeating scenarios played out in a boring plastic box. But New York artist **Paul Johnson** aims to set games free.

His surreal plastic-and-metal sculptural installations contain networked games that "play" each other. In *Maiden Flight*, the "client" computer assembles a virtual space station under the influence of metabolic data from the couch-potato star of the other game. *Crossings* holds two games in the same setting: a truck race through a countryside track where wildlife roam, and the the race from the point of view of the animals. The strangeness of the games extends to the hardware itself, a variety of components bolted to powder-coated metal racks and wrapped in plastic, courtesy of a vacuum former Johnson picked up on eBay.

"I've always taken apart disparate consumer technologies and rebuilt them to make something new," Johnson says. "It's the way I understand the creative process to work, bringing together different things and reworking them to make them my own."

The son of a computer programmer who worked on the Apollo space missions, Johnson has programmed games since he was a pre-teen. Previously known for his video art installations, he was recently drawn to gaming because it enables more interactivity between the elements. For example, *Dark Network* contains *Cruzaders*, a medieval skateboarding simulator with terrain that's constantly being modified by *M*, a puzzle of global commodities trading.

"The fact that *The Sims* can't network with *Quake* is interesting from an artistic perspective," he says. "You'd like to have these games from two completely different genres talk together. Traditionally, it's that kind of opposition of elements that create interest and conflict in a story."

These days, Johnson is honing his Linux chops to build PC clusters for future supercomputer-powered installations. "Artists should think of technology as another part of their creative toolkits," he says. —*David Pescovitz*

≫ **Paul Johnson:** pauljohnson.com

Grand Master Video

Audiovisual artist **Gardner Post**, whose pioneering work with Emergency Broadcast Network (EBN) inspired many of today's top VJs, is back with an epic new video performance tool called the Baby Grand Master, "The King of Video Instruments."

Billed as the Rolls-Royce of video performance instruments, the BGM features an impressive array of top-of-the-line audio and video technologies housed inside a baby grand piano cabinet, set atop a spinning turntable 8 feet in diameter.

Baby Grand Master is essentially a piece of sculpture that fits right in as a museum piece, yet is equally at home onstage as the ultimate live performance tool. The elegant white cabinet houses a pair of Pioneer DVJ-X1s, an Edirol V4 video mixer, and three Marshall LCD monitors.

An Allen & Heath Xone 92 audio mixer is accompanied by a rocking sound system, including dual 15" subwoofers, 18" and 12" subwoofers, and 3 bullet tweeters all powered by an AB 1,100-watt power amplifier. The piano's ultra-shiny white projection surface is perfect for blasting with video.

Post is offering a limited number of customizable units, with options that fit right in on *Pimp My Ride*. Perhaps you'd prefer your BGM with hydraulic legs and lid, or maybe neon under-lighting and running lights would do the trick. You can choose from any color of high-gloss lacquer finish, and there's also optional fog or laser assembly.

With their background and training as conceptual artists, EBN pioneered a new approach to working with video that now pervades mainstream culture. Their video remix of George H.W. Bush singing "We Will Rock You" to Saddam Hussein during the first Gulf War became a huge underground hit, and they left people speechless touring with the Lollapalooza music festival. Bono took notice, leading to their participation on U2's Zoo TV tour.

With this already enormous impact on today's audiovisual remix community, EBN founding member Post continues to inspire with his thought-provoking creations. —*Steve Nalepa*

≫ **Baby Grand Master:** babygrandmaster.com

Photograph by Till Krautkraemer

Photograph by John Marshall

Hungry Bots Must Eat

Ever seen a critter that moves like a spider, behaves like an ant, sees like a bat, and twitters like a bird? No one had, until Ohio State University assistant professor **Ken Rinaldo** built an army of ten robotic chimeras, each one capable of doing all those things.

The art and technology lecturer has long been interested in robotics that takes cues from the natural world. So it makes sense that Rinaldo had an epiphany when he heard entomologist Guy Théraulaz refer to ants as "rule-driven systems."

"I got excited because computers are rule-driven systems as well," Rinaldo says. "I thought, 'Wouldn't it be cool if they could feed themselves?' It would be solving one of the holy grails of robotics."

And so, in 2005, Rinaldo and talented helpers created the interactive installation *Autotelematic Spider Bots* for Britain's AV Festival. The group employed 3D modeling, rapid prototyping, and custom coding to produce six-legged robots that crawl around a rink in search of food and human interaction. The robots see using ultrasonic eyes, similar to bat sonar, on the end of their springy, antenna-like

necks. As pulses of energy bounce off spectators, the curious arachnids approach and engage, using sounds akin to sped-up bird chatter.

A real spider's gotta eat. So do Rinaldo's bots. They sense when they're hungry via a battery tester, and find food (a charger equipped with an infrared beacon) thanks to built-in sensors. To juice up, the robots again mimic biology: two rods out front dock with chargers, much like spider appendages, called *chelicerae*, bring food toward the mouth. As for the ant analogy, while the picnic invaders use pheremonal signals to broadcast where the fried chicken is, the bots use Bluetooth to pass on their paydirt's locale.

"I was interested to see if you could find a way to have a series of creatures that could communicate to the group, while also interacting with participants in the installation," Rinaldo says. "I have a core belief that natural systems represent the strongest models of what can work." —*Megan Mansell Williams*

≫ **Ken Rinaldo:** kenrinaldo.com

■◀ **Spider Bots in Action:** osu.edu/features/2006/rinaldo

ROUTER AESTHETICS

NOW THAT "DIGITAL CARPENTRY" HAS COME TO EXIST, HOW DO YOU MAKE IT AUTHENTIC?

By Bruce Sterling

THE "BLOBJECT" WAS THE DESIGN darling of the 1990s. Blob-shaped objects pushed the limits of what was technically possible: they married the fluidity of the new computer design to the amorphous qualities of high-performance plastic.

So everything from cellphones to major museums ached to be bulgy, bouncy, ripply, and radically curvilinear. The true joy of blobjects is that they're ergonomic — people are blobjects too, so when our intimate possessions become more forgiving and finger-friendly, we feel more at ease with them.

The 1990s loved blobjects in much the way that the 1930s loved streamlining: for many good and sensible reasons and some silly ones. There was always something a little uneasy-making about streamlined coffins and pencil sharpeners, and blobjects have limits, too. Not everything can aspire to the complex, luscious, plastic curves of a half-sucked popsicle.

At a glance, there's no way to tell if plastic will outperform cast iron or crumble like paraffin. It's hard to fully trust a swoopy Vernor Panton plastic chair: it looks like it ought to snap violently under your weight and sever your legs at the knees.

So how do you get the huge advantages of computer-aided design and machining without a big, expensive cauldron full of treacherous, colored goo? You could try a fabricator — also known as a 3D printer, stereolithographer, rapid prototyper, or rapid manufacturer. But, these futuristic gizmos aren't quite ready for the consumer prime time.

That leaves the humble router. The router doesn't spit goo like a plastics shop or layer stuff up like a fab; cheap and simple routers are the humblest of computer-controlled shop tools, basically just a spinning, toothy bit on metal tracks.

A router can dip up and down through the thickness of a sheet of plywood, and also roam from side to side across that board, slicing complex curves like a print head traveling across a sheet of paper.

The upshot of an afternoon's work with the router is a pile of crumbled router dust and a mess of complex 2D shapes, much like animal crackers. It remains to turn these shapes into something elegant and useful.

The Truss Collection is a commercially available set of chairs, benches, desks, and tables that all speak eloquently of their inherent routerliness. They were created by Scott Klinker of the Cranbrook Academy of Art.

> Scott Klinker is determined to make the router speak its own design language. A routered thing shouldn't be a mere downmarket knockoff of some earlier method of carpentry.

Like the other artists-in-residence at the Cranbrook Academy, Klinker works as he teaches. He and some Cranbrook students are determined to make the router speak its own design language. You might put it this way: now that "digital carpentry" has come to exist, how do you make it authentic? A routered thing shouldn't be a mere downmarket knockoff of some earlier method of carpentry. A router is a new thing in the world, so a clever designer should master it and use it expressively.

According to Klinker, there are three known methods of construction where the humble router naturally shines.

The first and simplest method is the "stack of sections." You take the routered pieces and pile 'em

Photograph by Bruce Sterling

**Digital carpentry: Unearthly elegance straight from the bargain basement.
A park bench created by Dharmesh Patel and Marty McElveen.**

up like a towering stack of pancakes. Of course, these flat pieces have to be glued, locked, nailed, or screwed in order to stick together, but this digital aesthetic can be striking. Instead of the sinuous, slimy, whiplash-line of a plastic blobject, stacked router sections convey a rugged, pixelly, computer-primitives look. They seem uniquely suitable for, say, the 8-bit penthouse pad of a playboy Super Mario.

Router technique two is the rather more sophisticated "graphic profiles." Rather than being piled and glued into a solid layered monolith, the flat routered pieces are turned on end and slotted together.

It's easy to cut jigsaw tabs and slots with routers, and this "graphic profile" treatment turns a router's stark cookie-cutterosity into a digital-aesthetic badge of pride. Tabbed and slotted structures of this kind have a Nipponese Superflat look, very angular and planar, perfect for the inhabitants of a pop-up book.

Then comes the last and most eye-boggling technique, the "fin model," aka the "grid of sections." This is a native router construction, that is pretty much nothing *but* slots and tabs: it transforms 2D into 3D by creating a reticulated honeycomb of intersecting wooden slices. The resulting constructions look finny and skeletal, like vector-graphic

simulations deprived of their skin.

One Cranbrook pièce de résistance was created by recent graduates Dharmesh Patel and Marty McElveen. Their weird green park bench bulges out of the Michigan forest soil like a cross between a half-deflated toadstool and an overstretched tennis net. Like the famous Wassily chair of Marcel Breuer, it's one of those far-fetched design fancies that turn out to be surprisingly comfortable.

A Wassily chair is a metal-and-leather designer skeleton that will cost you an arm and a leg. But a skeletal routered bench is a combo of software and plywood, two of the cheapest things in the modern world. It's cheap enough to abandon on the spot.

So, along with its aesthetic virtues, Patel and McElveen's bench is a futuristic advent that's also biodegradable. It's already sprouting spring weeds through its thousand open crevices, and melting slowly into the damp forest earth.

Bruce Sterling (bruces@well.com) is a science fiction writer and part-time design professor.

Hammer Time

Making the antiques of the future at Black Dog Forge.

By Kirsten Anderson

THE HIP BELLTOWN NEIGHBORHOOD of Seattle is an eclectic mix of old and new, where the last remnants of seedy bars and artist enclaves are uncomfortably tucked between new high-rise luxury condos and trendy restaurants. While the streets are becoming lined with high-priced boutiques, one particular alley of Belltown still retains its former bohemian allure.

Nestled amongst spray-painted murals and dumpsters is a purple door with a wrought iron gate. Above it perches a rusted metal skeleton pounding on an anvil. In the doorway a massive Rottweiler sits sentry, while the strains of Hank Williams' lonely croon float in the air, mixed with the clink and ring of metal on metal. You're at Black Dog Forge, where an ancient art is practiced in the middle of a city famous for cutting-edge technology.

The Forge, in its current incarnation, consists of blacksmiths Louie Raffloer and Mary Reid Gioia. Raffloer started the Forge in 1991. Originally a photographer, Raffloer was out one day drumming up work when he stumbled across a neighborhood blacksmith's shop. He was instantly hooked. Most blacksmiths would agree that it's an art that chooses you, rather than the other way around; Raffloer took a job in the shop, where he learned the foundations of his current profession. He eventually opened a few small forges before finally settling down with Black Dog.

Smithing goes back to ancient times, yet was almost rendered obsolete during the Depression as many big clients of blacksmiths went broke. The

Photograph by Kirsten Anderson

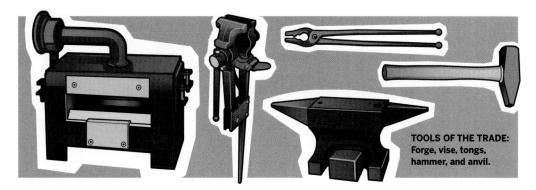

TOOLS OF THE TRADE: Forge, vise, tongs, hammer, and anvil.

rise of mass-produced objects pounded another machine-made nail into the coffin. After languishing for decades, blacksmithing began a national resurgence in the 1960s and 70s, when artisan-made objects became fashionable, and grew steadily through the 90s as clients (many of whom made their fortunes in the tech boom) began to get more money and a taste for exclusively made items.

Gioia picked up smithing much the way Raffloer did. Originally a custom upholsterer and sewer, she too first watched blacksmithing and became entranced. She wrangled herself some time at Black Dog, trading labor for time and instruction at the forge, and hasn't looked back since.

When asked about her initial attraction to working with metal rather than material, she laughs, "I liked that you could make really evil-looking stuff! Besides, it's sexy as hell." (Raffloer concurs, "It's a good conduit for a hyperactive ego because you have access to things other people can't do — it's romantic.")

Which isn't to say it's without a downside. "We bleed or burn on every job we do," says Gioia, although not without a teeny note of pride.

Since she began working with the Forge in 1993, Gioia's devotion to her craft has yielded a portfolio of work both functional and decorative, ranging from full bed frames entwined with sinuous black metal vines and leaves, to art-deco-meets-rock-and-roll iron necklaces, and, with Raffloer, even a huge iron railing for a castle in Mexico.

When discussing their training, both Gioia and Raffloer say the same thing: it's a job that's learned only by dedication and plain trial and error. "You really kind of teach yourself. Everyone learns as they go along and no one ever stops learning," says Raffloer.

A typical day at Black Dog might include a consultation with a client. "Obviously, handcrafted is more expensive than something at Restoration Hardware. But often people not only want the exclusivity of something handmade, they also enjoy the experience of working with an artisan to create something unique for their home," says Gioia. She might visit a client's house to gain insight into the client's tastes or to spy design elements that could be used in the commissioned piece. She then does three different sketches and lets the client choose one or decide if it should be modified.

Once the design is set, the actual forging of the piece starts. Materials are picked up, and often a template is drawn on a piece of wood to check for accuracy, especially if there are multiple elements that need to exactly match. In the case of, say, a curtain rod with decorative finials, Gioia also makes the brackets used for installation, and might use a jig (a forming tool) to ensure that measurements and angles are the same on each bracket.

Sound intriguing? Want to try it yourself? "In all honesty, you could set up a shop for under $1,000," says Raffloer. "Basically you need a hammer of any kind [Raffloer has been known to even bust out an occasional claw hammer to get the metal to do his bidding], a hunk of good steel over 100 pounds to act as an anvil, and a heat source." You can make a forge using firebrick, and Raffloer recommends using propane or natural gas to keep the fires burning: "Coal is romantic but impractical."

Any other advice for the budding blacksmith? "Yeah, get a significant other with a good job!" says Gioia jokingly. "There is a big learning curve, but it could be a good hobby, especially if you already have experience with metalworking."

Perhaps the most rewarding aspect, beyond creating the work itself, is the idea that it will exist long after you do. Gioia says, "Theoretically, everything I make could be around for thousands of years. I'm making the antiques of the future."

» Black Dog Forge: blackdogforge.com

Kirsten Anderson owns and runs Roq la Rue Gallery in Seattle. She spends her free time researching art, life science, and high weirdness.

Illustration by Damien Scogin

Maker

Photograph at top by Kirsten Anderson; below by Mark Frauenfelder

"We bleed or burn on every job we do."

OPPOSITE PAGE
TOP: Louie Raffloer
BOTTOM: Shop tools

THIS PAGE (CLOCKWISE FROM TOP): Mary Reid Gioia at work; the skeleton that presides over the entrance to Black Dog Forge; a warning sign; Duke, the shop dog

UNATTENDED CHILDREN WILL BE SOLD AS SLAVES

Big Blowhards

Who will be the first to make a machine that propels
a pumpkin more than a mile? By William Gurstelle

WHAT WEIGHS 28,000 POUNDS,
stretches 100 feet into the air, and shoots
vegetables, frozen turkeys, and bowling
balls? The biggest, baddest air gun of them all,
dubbed Second Amendment. This is the machine
that won first place in the 2006 World Championship
Punkin Chunkin, as it did in 2002 and 2003.

Built by a team of arc-welding air gun builders
headquartered in the exurbs of Detroit, Second
Amendment is an enormous, truck-mounted breech-
loader, a howitzer capable of shooting a 10-pound
pumpkin just shy of a mile. Last year, on the weekend
following Halloween, Second Amendment and a
couple dozen of its high-pressure cousins met in a
harvested cornfield in the wilds of Delaware to show
off their prowess at making stuff fly.

Punkin Chunkin (punkinchunkin.com) began
officially 21 years earlier, when its three founders
met informally to build hurling machines capable
of flinging leftover Halloween pumpkins. Little by
little, the hurlers improved their machines, and
every year the pumpkins flew a bit farther. Things
changed radically in 1995 when Trey Melson, one
of the co-founders of the event, upped the ante.

Jaws dropped and eyes widened when he
hauled Universal Soldier to the firing line. The
first truly giant pumpkin-shooting air cannon,
Universal Soldier ushered in a new era in hurling.
Now there are more than two dozen monster air
cannons, all capable of shooting projectiles more
than 2,000 feet. And a couple have even fired
5,000 feet.

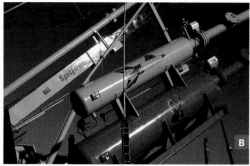

SHOOTING WITH AIR

The pumpkin guns of Delaware are simply the latest incarnation of machines capable of shooting things using air power, which is truly an ancient concept. Simple blowguns were used by prehistoric hunters to bring down small game. There are many references to breath-powered shooters by classical Roman and Greek historians. But mechanical (that is non-lung-powered) machines have a more recent history.

The oldest existing air gun is in the collection of the Livrustkammaren Museum in Stockholm, Sweden. The museum dates the device to around 1580 A.D. Air gun refinement occurred continually during the following 200 years. By the year 1800, air guns had developed to the point where they were likely more accurate and more powerful than contemporary black powder weapons of similar size.

For those who could afford them, air guns offered definite advantages: they were quiet and didn't produce target-obscuring smoke. But the main perk was that they could be fired rapidly — several times a minute, which was far quicker than the load, tamp, and fire procedure required of muskets. By comparison with smooth-bore, muzzle-loading muskets, air guns were veritable machine guns.

One of the most historically important American weapons was the air rifle carried by Meriwether Lewis during the Lewis and Clark expedition of 1803–1806. The actual gun (there is some controversy regarding its exact patrimony) may reside in the Virginia Military Institute's (VMI) museum near the stuffed remains of Little Sorrel, Stonewall Jackson's Confederate warhorse. VMI's .31 caliber, flintlock-style, 900 pounds-per-square-inch pneumatic rifle was crafted by expert clockmaker Isaiah Lukens in Philadelphia.

When fired, the Lukens rifle makes a weird, loud, whooshing sound instead of a bang. But it's powerful; it could easily take down a full-grown deer. According to the records kept by the Corps of Discovery, the air gun did its job well, being effective for hunting game and impressing enemies.

Air gun popularity waxed and waned throughout the next 200 years. In the late 19th century, England was swept by the "air cane" craze. Although an air cane appeared to be but a simple walking stick, inside it was everything needed to shoot a bullet with enough stopping power to take down a large attacker. Truly, the air cane was a dangerous weapon and an impressive means of self-defense for the security-minded Victorian Englishman.

In 1950s America, air guns were again peaking in popularity. As anybody who has ever seen the movie *A Christmas Story* knows, Ralphie Parker, and most boys like him growing up in the 50s, longed for a Red Ryder BB gun, or as Ralphie put it, "the Holy Grail of Christmas gifts."

Today, interest in air gunnery is peaking again, its health evidenced by a strong interest in airsoft guns, air-powered spud cannons, and — on the elephantine end of things — pumpkin guns.

PUNKIN CHUNKIN

At the 2006 Punkin Chunkin, held near Millsboro, Del., the long-standing distance records for pumpkin throwing came under serious pressure.

The current hot button in pumpkin chucking is breaking the mile barrier. A mile is a really long way to throw anything. Throwing a somewhat-fragile ten-pound pumpkin 10–12 city blocks is a task that requires simultaneous application of great power and extreme delicacy. Sure, tossing a cannonball such a distance might be small potatoes for a military cannon powered by 30 pounds of explosive cordite, but for the homemade air guns on the firing line, it's a stretch goal. Still, the people who compete here are clever and hardworking, and many of them seem to think they have a shot at the mile benchmark.

Currently, the state of the art in pumpkin hurling technology leaves the gun teams only a few hundred

---

Fig. A: **Even the spectators come prepared for the competition.** Fig. B: **Fire and Ice is a double-barreled pumpkin cannon.** Fig. C: **An entrant in the youth air cannon division, Soldier Too was built by a son of one of the founders.**

feet short of the magic mile mark. The 20,000-plus people attending the Punkin Chunkin love to watch the big cannons at work, for there's something undeniably interesting about a mile-shooting pumpkin gun made, for the most part, by regular Joes working in their garages and driveways.

Each gun is designed differently, and each has its own personality. Big 10-Inch is an engineer-designed gun that hides its top-secret workings behind an opaque covering. There are long, skinny cannons encased in spidery metal superstructures, such as Skybuster, Fire and Ice, and Please Release Me. There is the Harley-vibed Bad Hair Day, a gun crewed by a leather-clad female team, and the venerable and patriotically themed Old Glory, a former world record-holder that is always a favorite with the local crowd.

The technology behind these big blasters is not complicated, more a matter of scale than of high-

The Harley-vibed Bad Hair Day is a gun crewed by a leather-clad female team; the venerable Old Glory is a former world record-holder that's always a favorite.

tech engineering. The barrel of an air cannon is simply a large-diameter steel or aluminum pipe, usually scrounged from an industrial or agricultural scrap heap. The door and pumpkin-holding area inside the cannon are called the breech. Often this is simply a bolt-on steel plate near the bottom of the barrel and a grate on which to hold the pumpkin in place.

The basic principle of air- or gas-powered cannon operation is simple: rapidly introduce a powerful push of high-pressure gas from nearby storage tanks into the breech of the cannon and, in doing so, push the pumpkin out of the barrel as forcefully as possible. Valves control the release of the high-pressure air. The valve design is critical: a too-small valve, even if it opens very quickly, will retard the accumulation of pressure inside the gun and hinder performance. A big valve that opens too slowly will do the same. The most resourceful builders incorporate a big,

fast-acting group of valves that instantly open the floodgate of high-pressure air or gas.

This competition isn't about money or trophies or even the teardrop-shaped mass of gooey pumpkin flesh a mile away in the middle of a Delaware farmer's field. It's about pride. It's about being the best, about setting a goal and achieving it. When told of the time and money invested in making these guns, a lot of people simply smile and shake their heads. But not real makers. Real makers understand.

▣ See the Maker File video episodes from Punkin Chunkin: makezine.com/blog/archive/the_maker_file.

William Gurstelle is a MAKE contributing editor. He wrote "Happy Blastoff" for MAKE, Volume 10.

Fig. D: Centrifugal hurler Bad to the Bone won its category with a throw of 2,737 feet. Fig. E: The contest's 10-inch white pumpkins are hard and gourd-like. Even so, some pumpkins explode or "pie" as they leave the cannon. Fig. F: A pumpkin emerges out of smoke from the barrel of Old Glory, which fired 3,632 feet for third place. Fig. G: The wicked-looking Pumpkin Slayer catapult finished second. Fig. H: This catapult used a bicycle to draw back garage door springs. Fig. I: John Buchele's bright red Great Emancipator cost him $70,000 to build.

The Maker State

Safe working practices give you the freedom to attempt projects on the edge. By William Gurstelle

WHEN MY SON ANDY WAS 12 YEARS old, he entered his junior high school science contest. His challenge was to invent something new and useful. He badly wanted to win, but inventing something useful is hard, especially when you're 12.

After various aborted attempts, he came up with a self-buttering toaster. What a brilliant idea from such a young person! (I readily admit my fatherly bias.) The device was intricate yet simple: a wood and steel construction that held a slice of bread at an angle in front of a carefully wound matrix of nichrome wire heating elements. While the bread toasted, the heat from the wires melted a glop of butter on a perforated metal holder positioned over the bread. The butter dripped through the holes

and onto the toasting bread. Voilà! There was a slice of automatically buttered toast. By my lights, this was a pretty terrific invention for a sixth-grader.

The evening of the fair approached, and Andy and I looked forward to a night of glory. The judges, a collection of teachers and parent volunteers, methodically walked up and down each aisle. They asked questions, measured things with rulers, made notes on clipboards, and generally maintained a judge-like demeanor. When the judges came to Andy's table, the toaster worked perfectly. With self-assurance and a smile, he handed each judge a slice of warm, buttery Wonder Bread for a snack.

But when the winners were announced, Andy's name wasn't called. Crestfallen, he approached the judges and asked, "Why didn't I get a ribbon?"

"Well, Andy," said a judge, "we thought your machine was dangerous. After all, it uses electricity and it gets very hot."

"Of course it does. It's a toaster," he protested. "It's supposed to get hot and use electricity. If it didn't, it wouldn't be a toaster." Unswayed by logic, the judges would not reconsider.

So who won? First place went to a girl who made a cap and vest for her hamster. Second place went to a boy who "made" radar by cutting out pictures of antennas and gluing them to a poster board.

Some might say our society has become obsessed with safety to an unhealthy degree. There are labels that say "do not use in shower" on hair dryers, "do not eat toner" on laser printer cartridges, and "allow to cool before applying to groin area" on McDonald's coffee cups. (The only one I made up is the last one.) Some people feel that everyone else ought to watch out for them. They want someone to vet their lives, to check things out in advance. Basically, they want a nanny.

In a "nanny state," somebody else — governments, insurance companies, education administrators — decides which projects makers may attempt and which they may not. In the nanny state, experimenters and builders find themselves deprived of the materials, tools, and information they need to carry on their interests.

On the opposite end of the spectrum is the "night watchman state." Here, authorities try to keep thugs off the street, keep the electricity on, and that's about it. You're pretty much on your own.

Most of us prefer to live, work, and play somewhere in the middle. Let's call it the "Maker State." In the Maker State, everyone takes reasonable precautions and wears protective equipment. Safe working practices, if thoughtfully incorporated into the act of making things, can become a performance-improving feature, just as athletes wear better equipment to enhance their performance.

The Maker State provides freedom to attempt projects on the edge. Still, laws of chance and statistics ensure that sometimes stuff just happens. There are two fundamental realities of working in the Maker State: risks can be reduced but not eliminated, and not everything is somebody's fault.

It's up to each person to determine his or her personal "zone of reasonableness." It's not the same for everyone. It depends on the quality of your equipment, the extent of your experience and training, and your willingness to assume risk and responsibility for your actions.

In the late 1950s, Americans, especially teenage boys, went rocket-crazy, their interest lifted to stratospheric levels by the patriotic frenzy surrounding the launch of the Soviet Sputnik satellite. By the early 1960s, thousands of young people were busily building homemade rockets.

Unfortunately, few had any idea what they were doing, so most wound up building what in reality

There are labels that say "do not use in shower" on hair dryers, "do not eat toner" on laser printer cartridges.

were pipe bombs. Unarguably, mixing inexperience, surplus enthusiasm, and powerful chemicals makes for a dangerous situation.

Estes Industries, now the biggest name in manufactured model rocket engines, published a booklet 40 years ago called *The Rocketeer's Guide to Avoiding Suicide*. It provided example after chilling example of rocket engine explosion injuries, some presented in gruesome detail, e.g., "He was making rockets out of pipes filled with match heads. The pipe blew up and he almost blew his stomach and intestines out."

While Estes had a vested interest in persuading young rocketeers to buy their engines instead of building them from scratch, nonetheless, there did seem to be an extraordinarily high accident rate among youthful rocket builders of the time.

Makers of the 1960s might not have known about various hazards that we now recognize. We can't ignore these hazards. Instead, we must learn how to avoid them and work safely by taking precautions and wearing protective equipment. By employing adequate preparation and knowledge, and by incorporating safety as a positive, performance-enhancing feature in projects, the Maker State engenders a wide variety of challenging projects. With any luck, the next self-buttering toaster at the science fair will win the blue ribbon.

William Gurstelle is a MAKE contributing editor. He wrote "Happy Blastoff" in MAKE, Volume 10.

Drawbot Love

How a bunch of us made our own drawing robot. By Bre Pettis

WHEN I FIRST READ ABOUT DOUGLAS McDonald's Scribbler Bot (*MAKE, Volume 07, page 141*), it was love at first sight. I simply *had* to make a drawing robot.

Doug's original Scribbler Bot converted webcam photos into distinctive line drawings, then used a homemade plotter (with a pen or pencil zip-tied on) to render them onto poster-size paper. I knew from his article that to put something like this together myself, I needed to get some stepper motors and boss them around with some software. Luckily, I got a lot of the hardware issues out of the way by finding a Japanese medical contraption that had a former life organizing vials of blood. It was a perfect XYZ platform for my drawbot!

I quickly realized that I couldn't do this project on my own. The hardware required reverse-engineering, and the software had to be coded.

My friend 3ric held a robot-making get-together at Seattle's Public N3rd Area, and friends were recruited to help. Fueled by undocumented quantities of pizza and Mountain Dew, contributors 3ric, Adam, Melvin, Brian, Divide, John, and Choong brought their ninja-level hardware-hacking and software-writing talents to the project.

On the hardware side, we hooked up the steppers and the limit switches to the MAKE Controller, and we put together DB9-connected serial cables with different-colored wires so they would be easy to follow if there was a problem. I found that when running lots of wires, it helps to twist them all up into a cable with a drill, and when attaching them to things, zip ties are your friend.

Throughout the build, it was important to keep a notebook with all of our diagrams and notes. The stepper motors required more power than the

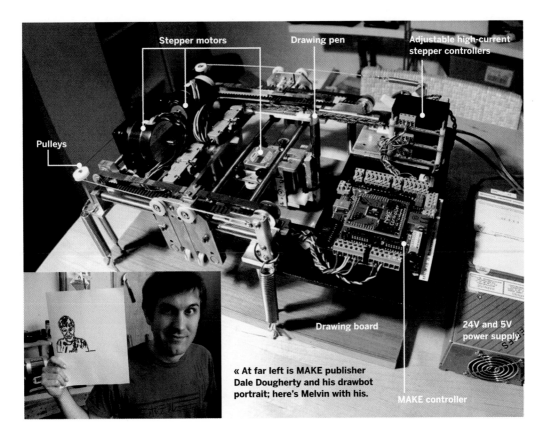

Stepper motors

Drawing pen

Adjustable high-current stepper controllers

Pulleys

Drawing board

24V and 5V power supply

« At far left is MAKE publisher Dale Dougherty and his drawbot portrait; here's Melvin with his.

MAKE controller

MAKE Controller could put out, so I ordered some Interinar microstepping motor controllers that could be adjusted to output the power the steppers needed.

Holding the paper down turned out to be somewhat tricky — we needed a separate base and springs to hold it stable. We added legs to the contraption, and John Blunt, our woodworking neighbor, made a beautiful oak base with clipboard clips to keep the drawing paper secure.

The drawbot process starts with a photo taken by my MacBook Pro's iSight camera. Any image would work, but using the iSight removes the inconvenient step of importing photos to the computer. Then you save the image as a *.bmp* file, and drop the file into our Launch Drawbot program. Launch Drawbot converts the color image into a simple black-and-white bitmap using Peter Selinger's mkbitmap utility, and then converts the resulting bitmap into a vector graphic representation using Selinger's Potrace. Mkbitmap and Potrace are both open source, available on sourceforge.net.

Launch Drawbot shows you a preview of the drawing before you start, so you can get an idea of how it will work. You can also adjust the size of the dark areas, where the contrast edges are drawn,

and how thick the fill lines are. The better the image going in, the better the drawing coming out will be, and we discovered that filtering the image before generating the vectors is critical to reducing the line count, which reduces drawing time. We didn't want to wait 8 hours for our pictures.

Once the actual drawing starts, the program sends packets of data over Ethernet to tell the drawbot which coordinates to go to. As soon as you command the drawbot to begin, it puts the pen down on the paper and starts drawing. It draws an outline of all the areas first, and then goes back and fills in the shading.

Everybody who worked on the drawbot agreed that no matter how much you suffer from OCD, it's spellbinding to watch and can maintain your attention for hours on end. Feel free to download the code for the project, play with it, and make it better. It's under the GPL license, which means you're free to use it as long as you release your changes under the same.

➕ For more, go to makezine.com/11/drawbot.

Bre Pettis produces Weekend Projects and other MAKE Video Podcasts on the MAKE blog at makezine.com/blog.

COOKING WITH TRASH

By Tim Anderson

The Chinese make fuel briquettes from coal dust, farm waste, scrap wood, and a bit of local red dirt.

Known for their ingenious reusing and wasting not, the Chinese make very effective cooking fuel briquettes from waste cellulose and carbon materials, using local clay as a binder. The briquettes are cylindrical with air passages through them.

An expat living in the United States told me he had a water heater in Germany in the 1960s that used the same type of briquettes.

The source materials for the fuel include coal dust, carbon from sawdust, farm waste, and scrap wood. The local red clay dirt is the binder.

HOW THEY DO IT

Sawdust and other carbon-bearing materials such as waste paper are first reduced to charcoal in a furnace. I haven't yet seen this part of the process.

Coal dust is used as is. The coal is ground to a certain size and mixed with the red clay dirt. The ratio is 80 percent coal to 20 percent clay.

At the same time, the mixture is sprinkled with water from a watering can; enough water is mixed in so that it will all hold together when squeezed by hand. A certain amount of sand or limestone gravel is present in the dirt and doesn't seem to cause a problem. That's it!

When completely burned, a "clinker" is left over, which is removed with tongs. Stomping on them easily crushes these clinkers.

They are then often used as what seems to be a decent road surface, as well as good soil for growing crops.

Videos of the briquette maker in action:
makezine.com/go/briquette1
makezine.com/go/briquette2

Tim Anderson, founder of Z Corp., has a home at mit.edu/robot.

A, B, C: Multiple views of the machine used to produce the briquettes. D: A food vendor's bicycle rickshaw with briquette burner. E: Top view of the burner, showing the brick lining. F: Trash and compost in the garden; a squash plant fertilized by "clinkers."

Power Tripping

High-voltage engineer Greg Leyh builds the largest Tesla coils in the world.

By David Pescovitz
Photograph by Jonathan Sprague

GREG LEYH IS A POWER BROKER. VOLTS.
Watts. Joules. Amperes. He trades in them all. His aim? Nothing short of lightning-on-demand.

At Maker Faire in May, Leyh premiered his twin Tesla coils, two stately and elegant 10-foot towers that spewed 18-foot arcs between them. Amazingly, the twins are just one-twelfth-scale prototypes for the pair he plans to build at his Nevada Lightning Laboratory. Those coils will fill a football field-sized tract of land with 18 million volts of lightning.

For Leyh, high voltage is a way of life. His company, based just south of San Francisco in the small city of Brisbane, is called Lightning On Demand. In a few years, LOD (lod.org) will be reborn as the Nevada Lightning Laboratory, where, if all goes as planned, he'll open a world-class facility for scientists to study high-power phenomena.

"The higher power you go, the more new physics you uncover," says Leyh, who works days in the Power Conversion Department engineering group at the Stanford Linear Accelerator Center. His personality somehow fits his job. He's quiet, very friendly, a little nerdy, and always willing to explain technical concepts repeatedly until you understand them, or think you understand them. In that way, Leyh reminds me of the best high school science teacher, the kind who still dresses like a NASA engineer from the 1960s — short-sleeve dress shirt, pens in his breast pocket, plain slacks, and dress shoes. But instead of a slide rule on his belt, Leyh wears a calculator watch on his wrist.

If the Nevada Lightning Laboratory can collect just $12 to $18 million in funding, Leyh says he could generate the first arcs in little more than two years. He's just returned from visiting a site 40 minutes outside of Las Vegas that, based on his meticulous surveying, GPS mapping, and Google Earth exploring, would be the perfect spot. Now he just has to finalize a deal with the governmental owners of the property.

Leyh first put his ideas for the Lightning Lab on paper in 1996, but he's been on a power trip since his teens. As a science-minded high school senior in Arlington, Texas, he stumbled upon proto-maker Nikola Tesla's writings about resonance rise, the phenomenon that causes a street light pole to sway wildly at the top from just a small shove at the bottom.

"The whole notion that these physics effects are not only knowable, but can be calculated very precisely, was almost too much to believe at first," Leyh says.

As a college freshman at UT Arlington doing work-study in the machine shop, Leyh decided to conduct his own mechanical resonance experiments. He built a mechanical oscillator from an old Camaro's blower motor. Essentially, the machine repeatedly lifted itself up and then dropped back to the ground, at various speeds controlled by a rheostat. Leyh attached it to various objects to determine their resonant frequencies. His most, er, successful field study took place on a wooden footbridge.

"I adjusted the dial until I found the sweet spot where the bridge was bouncing a foot and a half up and down," he says. "Then I heard a very satisfying crack and I couldn't find the right frequency again."

Through college, Leyh devoured Tesla's writings, eventually building his first small Tesla coil. The coils also exploit resonant rise, but with electrical energy rather than mechanical. A Tesla coil steps up the power from an input source by taking it through several transformer and driver circuits until it reaches incredibly high voltages. That energy is then discharged in zaps of radio frequency (RF) energy.

After graduating with an electrical engineering degree, Leyh landed a job working in Stanford's physics department. In 1988, a friend sent him a grainy, fourth-generation video of San Francisco machine performance group Survival Research

Leyh built the twin Tesla coils in the garage of his Brisbane, Calif., home, a melding of modernism and industrial beauty that he designed and built almost entirely himself.

See more photos of Leyh's workshop and his Tesla coils in action at makezine.com/11/proto

Laboratories. Intrigued, Leyh eventually located SRL director Mark Pauline's legendary machine shop and the two became fast friends and collaborators.

Shortly after his first meeting of the minds with Pauline, Leyh noticed that the scraps from a decommissioned particle accelerator at Stanford were headed for the dumpster. Digging around in the big-science garbage, Leyh pulled out transformers, copper stock, basically "80 percent of the makings of a big Tesla coil." He hauled the detritus to the SRL shop and set to work. In 1990, Leyh's 40,000 Watt Experimental Coil made its debut at an SRL performance in Seattle. At the time, it was the largest Tesla coil in the world.

"The Lorentz Gun creates a 40-foot section of a real lightning bolt."

Leyh and Pauline also developed several other machines, the most infamous being the 110,000-volt Lorentz Gun, formerly known as the Taser. The name change was spurred by a nastygram sent to LOD several months ago from TASER International. The company claimed copyright infringement even though Leyh built his device years before TASER trademarked the word.

The guts of the Lorentz Gun are 4,000 pounds of capacitors that SRL intercepted on their way from Lawrence Livermore National Laboratory to a waste management facility. Wired together in a bank, the capacitors produce 110kV at 100 kilojoules, enough to blow a big divot out of thick steel. The gun fires a thin wire into the target, and the capacitors are instantly discharged. Within the first 100 microseconds, the wire melts into plasma. Even with the wire gone, the current is contained within a magnetic field and delivered to the target.

"The Lorentz Gun essentially creates a 40-foot section of a real lightning bolt," Leyh explains.

As SRL featured both the Lorentz Gun and the Tesla coil in shows, Leyh's reputation grew as a high-power researcher with, well, unusual application ideas.

In 1996, Leyh met Eric Orr, a Los Angeles artist known for large sculptures involving fire and water.

A wealthy art patron in New Zealand had commissioned Orr to build a "fountain for lightning." Generating lightning was Leyh's specialty, so a mutual friend introduced the two. Two years later, Leyh, Orr, and a team of assistants, mostly from SRL, set up camp on the patron's property and installed *Electrum*. The 38-foot-tall sculpture generates 40- to 50-foot lightning bolts against a sweeping backdrop of coastal waters. To this day, it is the largest Tesla coil in the world. Leyh has never been called back to New Zealand for maintenance.

While Leyh's projects have charged crowds around the world, to him they're all incremental steps toward the Nevada Lightning Laboratory. He conceived of the facility a decade ago when testing new simulation software for the Stanford accelerator. To put the software through its paces, Leyh built a virtual Tesla coil. And then a bigger one. And so on.

"I found that two 120-feet-high coils operating in opposite phase is right before the point of diminishing returns," he says. "So it follows logically that a facility of that scale should exist."

The coils will open a window onto uncharted areas of mega-scale electrical physics, providing scientists with the opportunity to get up-close and personal with lightning. For example, Leyh expects research groups will value fresh data that may deepen our understanding of how lightning is related to global temperature changes. Yet while the business plan counts on research dollars, the bulk of the financial support is expected to come from tickets to public educational demonstrations, some of them seen from the top of one of the coils.

Each 12-story-high tower will be topped with a 55-foot-diameter electrode. Amazingly, the operator and observers will be inside the electrode. According to Leyh, "That's actually the safest place to be." If visitors get nervous, they can always sidle up to the bar that Leyh designed for the east tower. The west tower's electrode will include living quarters for experimenters or, Leyh points out, a donor's apartment.

Right now, though, the Lightning Lab exists only on paper and in Leyh's hard drive. "It'll never be done. Once it's constructed, there will be an endless stream of questions that need to be answered," Leyh says. "Anytime you devise a truly new scientific instrument, you find things you could never have imagined."

MAKE Editor-at-Large David Pescovitz is co-editor of boingboing.net and a researcher at the Institute for the Future.

Leyh takes measurements from inside *Electrum* before it's shipped to New Zealand. Alberta Chu's film *Electrum*, shown on PBS, documents the project. asklabs.com/electrum

LONG LIVE THE BICYCLE

EVERYTHING I KNOW I LEARNED FROM TWO WHEELS AND A FRAME.

By Saul Griffith

I REMEMBER THE VERY FIRST MOMENT I rode a bicycle. I was at Uncle Dave's place out "in the bush." He wasn't a real uncle, but rather one of those family friends who becomes a default uncle by giving freely of his time and lessons on life. He lived a two- or three-hour drive from Sydney in a small town, on a beautiful rustic property with a shed full of the things that delight a 6-year-old, and one of those things was a bicycle.

There were no luxurious training wheels, just two tireless men — my father and Uncle Dave — who would run behind me holding the underside of the saddle and keeping me upright as I teetered and tottered. It took an afternoon, a beautiful afternoon of giggling and grazed knees, but I was anointed a bicycle rider, and was then allowed to ride to the edge of the property and back. It was probably only a hundred yards, but the world suddenly seemed larger. The love affair would never end.

Soon after, I received a purple chopper, with a banana seat replete with metallic flake. (My sister's banana seat had flowers on it, something I derided at the time, but now appreciate and even search for in old bicycle shops. To my mind, a chopper isn't complete without one.) At age 7 or so, inspired by the sublime acting of Nicole Kidman — I kid you not — in *BMX Bandits*, I convinced my parents to upgrade it and fulfill my dream of owning a black and yellow Speedwell. (It had to be black and yellow.) Not long after that, *E.T.* came out at the movies, and my fantasies of flying bicycles powered by my own little alien were frequent.

What really came with bike ownership, though, was bicycle maintenance, my first taste of hands-on engineering. When I got my first mountain bike,

a coral pink Apollo brand, I would tear it down and build it back up just for kicks. The bearings and their hardened steel balls fascinated and perplexed me as I cleaned, greased, and serviced them. I learnt the difference between left- and right-handed threaded screws by cross-threading them. I learnt about galvanic corrosion by riding on the beach with steel pedals and aluminum cranks. I learnt about gears, chains, derailleurs, and broken teeth. (The broken teeth — the gears' and my own — were the result of not correctly tensioning the chains. Fortunately, mine were baby teeth and my engineering prowess improved in time for the onset of adult teeth.)

For many of us, bicycles are the first taste of responsibility. As soon as you start modifying or repairing your own bike, you learn very tangibly the results of your work. If you do sloppy work, or make mistakes, the result is typically a bicycle crash — blood, broken bones, and all.

But bicycles taught me more than just basic building principles and simple mechanics; they even introduced me to the magic of materials science. After I tired of my bike's pink color (the only one available), I decided that chrome would be far more "manly." I saved enough money to strip the bike down and take it to a chrome plater.

I clearly remember going to the factory with my tolerant father and being fascinated with the electroplating baths, although the whim turned out to be disastrous. The process effectively annealed and weakened the frame, and the fork gave out soon after, bending slowly upward until it was unrideable. The experience was probably influential in my eventual university study of metallurgy — the fatigue and

This plastic bicycle was a whimsical design exercise in making a complex 3D object from flat 2D sheets of poly-carbonate. It's heavy and cumbersome, and rides like a wet noodle, but Wonder Woman wouldn't be caught riding anything else.

Photograph by Saul Griffith

properties of metals were very obviously important to the practice of engineering.

Riding my bike taught me about the physical universe. For a period, my friends and I irrationally thought it uncool to have brakes (not that brakes were very good in those days anyway), so we would remove the brakes and use the soles of our shoes rubbed directly on the rear wheel as the slowing mechanism. From this, I quickly learnt about friction, heat, and energy dissipation. I remember multiple trips to the shoe store with a disgruntled mother intrigued as to how I could wear out my shoes in only a month with such an unusual wear pattern.

My friends and I worshipped the older boys in the neighborhood who had fancier bicycles and could make them do the impossible, like doing a "wheelie" — riding on the rear wheel only — from one end of the street to the other. Then came the fantasy of the "bunny hop" — jumping the bicycle without a ramp or gutter. Only years later would I learn that this is a non-trivial trick of physics and requires timing the movement of the center of mass carefully.

In high school, my favorite class was technical drawing. We did the high school equivalent of basic structural analysis, and learned drafting skills with pencil, paper, drawing board, and a bevy of protractors, compasses, and guides. Naturally, at the first open project we had, I labored for weeks with the design of a bicycle. Learning geometric principles with passion, I drew complete technical specifications and what would be my first industrial design rendering. (I like to fancy that the work of my imagi-

nation was a precursor to the carbon fiber mono-coque frames that soon after became the fashion in Olympic track cycling.) I would ride endlessly around the velodrome in our neighborhood, and along with my friends, I would fantasize of sporting glory while learning the principles of centripetal forces on the sloped oval track.

I may not have gone to the Olympics, but to this day I still build my own bicycles, whether from the cherry components I've always dreamed of, or by starting from scratch. I've made bicycles entirely out of plastic, built wooden bicycles, and even designed a bicycle "Lego" kit of basic frame components that can be reconfigured into all sorts of different geometries. Bikes haven't stopped teaching me new things: I first welded on a bike frame, and I learnt much about energy efficiency by analyzing bicycles as a transportation alternative. (If you can do only one thing for the environment this year, let it be giving up your car for the bike, or at least picking up your bike helmet instead of your keys a few times a week.)

What more can I say? I love bicycles. I love the feeling of speeding silently under my own power. Maybe what I'd like to say is that, more than love, I owe the bicycle. I owe it a debt of gratitude. As a growing engineer, it encompassed nearly every principle of science and engineering that fascinated me. Long live the bicycle! Long live those who tinker with them, even when they make mistakes.

Saul Griffith works with the power nerds at Squid Labs.

Ten-Second Stomp Rocket
By Emma Wagstaff

Make a juice-bag-powered rocket in less time than it takes to drink it.

You will need: An empty Capri Sun (or other foil-lined juice bag with a straw), a marker, some tape, and a foot

Make it.

When my dad and I became frustrated building a complicated model rocket, we took a snack break and I had a Capri Sun. After I drank the juice bag dry, I figured out that I could make a stomp rocket in about ten seconds. I folded over one end of the straw that came with the juice and taped it closed. Then I blew air into the bag to inflate it and inserted the straw back into the hole. I put it on the floor, and "5, 4, 3, 2, 1 … blast off!" I stomped on the bag and the straw shot across the room. It was way faster than building our rocket and just as cool.

TIP: It helps to mark the straw about 1" from the insertion end, so you don't put it in too far. This prevents the stomping foot from stepping on the straw rocket.

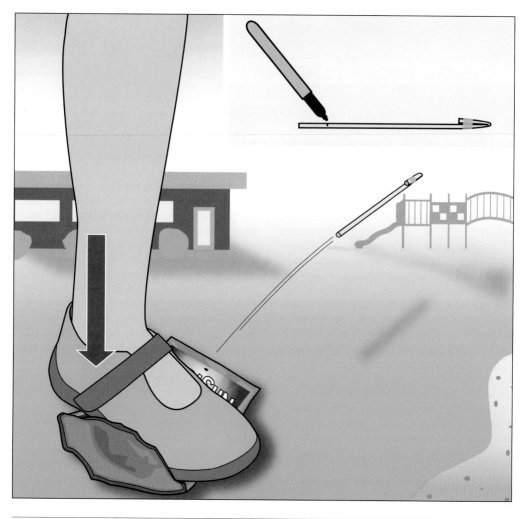

Emma Wagstaff, now age 9, is a fourth-grader at Ohlone Elementary School in Palo Alto, Calif. She loves sports, including soccer, baseball, surfing, and windsurfing, and she loves reading and doing all kinds of projects.

Illustration by Tim Lillis

alt.vehicles

DIY WHEELS

Newer cars aren't maker friendly. They require special tools to maintain, and manufacturer's parts come in the form of outrageously expensive black-box modules. But older cars and bicycles are wonderfully hackable platforms. They're inexpensive and easy to customize. In this special section, we'll show you how to build your own chopper bike, mobile movie projector, bike iPod charger, and more. **Vroom!**

PEDAL-POWERED PICTURE SHOW:
Be the hit of your neighborhood.

Photograph by Robyn Twomey

Mister Jalopy's Urban Guerrilla Movie House

Your own DIY drive-in.

Rather than lamenting the slow death of drive-in theaters, I decided to build my own, mount it on an adult tricycle, and take the movies anywhere there's an AC outlet. Vibrant online communities of DIY projector enthusiasts have ironed out the kinks and built the focal calculator software tools, and they're building homebrew machines with jaw-dropping results.

The folks at Lumenlab (lumenlab.com) have put together a kit of hard-to-find components to build your own movie projector using a surplus 15" LCD computer monitor. Rather than scour the web for components that might work, I decided to take advantage of Lumenlab's engineered solution, their considerable informational build guides, and, most of all, their invaluable forums that include detailed project logs of completed projectors.

Mister Jalopy is a mediocre welder, a fair shade-tree mechanic, and a clumsy designer, and has never touched a piece of wood he hasn't ruined. However, he still gets a lot of love at hooptyrides.com.

MOBILE MOVIE MADNESS

Here's how your backyard fantasy drive-in really works.

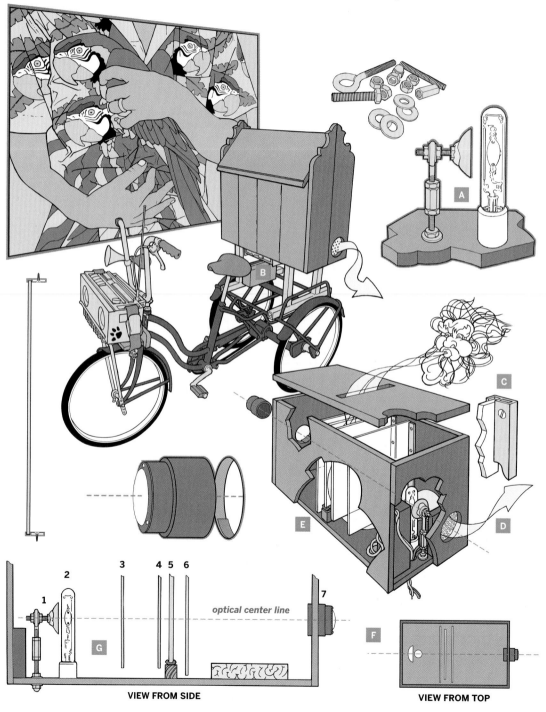

optical center line

1 2 3 4 5 6 7

VIEW FROM SIDE

VIEW FROM TOP

A look inside

A. The light source of the bulb is the capsule suspended inside the glass envelope, so center the reflector on the capsule, not the entire bulb assembly.

B. Two ballasts? No. Initially, I was going to mount the ballast inside, but since I had the space, I moved the ballast outside to help cool the projector.

C. An aluminum window screen channel holds the Fresnel, heat shield, and LCD upright in the cabinet. I smushed a small lump of epoxy putty in the channel as a stop to hold the Fresnels and heat shield at the right height. The LCD rests on a little block of wood.

D. Cool air enters the top of the cabinet and is drawn over the LCD by the fan at the rear. Heat is a big issue as LCDs don't like temperatures over 130°F, and the bulb is hot like the sun.

E. Ideally, the LCD would be perfectly sealed to the edges of the cabinet. In the real world, light leaks around the LCD and through cabinet cracks can be addressed with aluminum flashing tape and foam weatherstripping.

F. The complexity of the project is largely in placing the objects. Figuring out the center of the Fresnels or the triplet lens is simple — just measure and divide by two. Finding the center of the LCD is more tricky as you need to find the optical center not the physical center. For example, let's say the LCD assembly is 14" high from the top of the glass to the bottom of the cables. The physical center is at 7", but that includes a bunch of cabling that is not optically relevant. Everything needs to be centered off the portion that is optically important — the center of the LCD projected image, the light emitting portion of the bulb, and the glass reflector — not the whole base.

G. Based on the lens used and the screen size desired, the focal calculator software will help you position these objects in your projector.

How does it work?

An LCD computer monitor has a very even but fairly dim fluorescent backlight behind an LCD glass panel. If you remove the backlight and replace it with a powerful light source, an image is projected that can be captured with a few lenses and thrown up on your personal movie screen!

With a 400-watt metal halide bulb (Figure G, label 2) as the light source, a reflector (1) nudges the light past the heatshield (3) toward the first Fresnel (4) lens. Like an old-school overhead projector, a Fresnel (pronounced fray-*nel'*) lens is a plastic sheet with molded, concentric, ridged circles. The first Fresnel (220mm) acts as a condenser to provide an even light source to shine through the LCD panel (5).

The 317mm Fresnel lens (6) picks up the projected image that is then captured by the triplet lens (7) at the end of the projector. To make sense of how the projector works, I picture the light pushing the LCD image through the projector to be shot out the lens on the end like a sprinkler spraying the picture on the screen, like weightless water.

Though my projector looks crude, it's actually built to millimeter precision. While exact placement of the non-optical components like the heat shield and fan is not critical, the distances between the bulb, Fresnel lens, LCD, reflector, and triplet lens must be exact to the calculated dimensions.

My projector's first image was unbelievably blurry. If I hadn't allowed for adjustability of the Fresnels and triplet lens, the project would have gone down in flames and this article would have been about a talking rooster named Charlie.

MATERIALS

LUMENLAB MEGA-PROJECTOR KIT INCLUDES:
- 220mm and 317mm Fresnel lenses
- 320mm projection triplet lens
- 400W metal halide bulb with external high-voltage ballast
- Mogul-style porcelain socket for bulb
- 12V fan with BBQ-grill-style guard, silicone dampener, and AC adapter
- Reflector

Focal calculator software free at makezine.com/go/lumenlab

15" LCD display panel XGA 1024×768 resolution or higher, 15ms response time or lower, 0.28 dot pitch, 400:1 contrast ratio or higher

Latex gloves, anti-static mat, and anti-static wrist strap advisable for salvaging the fragile LCD screen

Aluminum channel from hardware store DIY window screen kit

3" ABS pipe and pipe coupler

⅜" bolt, nuts (2), and washers (2)

¼" eyelet bolt and nuts (2)

¼" threaded rod

¼" coupler nut

Epoxy putty

Masking tape

Sandpaper

Dust mask

4" hole saw

Velcro

Lexan plastic heatshield

THE BOX
I used a Schwinn Town and Country and wood from the scrap pile, a set of garage sale shelves, and a kitchen table from the trash.

Mocking up the components will allow you to picture the enclosing box you'll need to build. MDF or plywood would have been a much quicker path to guerrilla movies.

Illustration by Tom Parker

DIY DRIVE-IN FAQ

Which 15" panel is recommended?

Lots. I used an HP model 15VP, but check the Lumenlab database (makezine.com/go/lcddata) for considerable community experience with most panels. Since the fluorescent backlight won't be used, you can use a "broken" panel with a blown backlight.

What does "stripping the panel" mean?

To project light through the display panel, you need to strip the monitor to a sheet of LCD glass that you can see through. Besides removing the fluorescent backlight, you also need to swing all supporting circuitry out of the way.

What sort of problems might I encounter when stripping a monitor?

Since the LCD panel needs to be clear of obstructions, you may find that its FFCs (flexible flat cables) are too short to move supporting components out of the field of projection. The Lumenlab FFC extension cable may give you the needed extra distance, but unless you're dedicated to a particular unit, it's preferable to select one of the many monitors that don't have FFC issues.

How careful do I have to be when taking apart the LCD panel?

Super careful! The LCD glass is thin and manufacturers can be pretty tricky about hiding tabs and screws. (Didn't they read the Maker Bill of Rights from MAKE, Volume 04, page 157?)

Can I use an LCD from a laptop?

No. As a function of laptop supermega-compactness, the motherboard, video card, and monitor circuitry are so tightly bundled together that the monitor functionality is not easily separated from the laptop.

Can I use a car headlight, halogen work light, old overhead projector light source, etc.?

Probably not. There are a lot of factors to consider like color balance, brightness, type of light, and heat issues. But search the Lumenlab forums.

What should I use as a video source?

Whatever you want, as long as your selected donor monitor has the required input. Since I used a computer monitor with a VGA-style input, I used my DVD-playing laptop as a source.

What about old overhead projector panels, PSone screens, LCD TVs, my grandpappy's rear projection TV, etc.?

Probably not, as they have disappointingly low resolution, slow refresh, incorrect inputs, and other shortcomings. Again, search the Lumenlab forums.

⚠️ **CAUTION! CAREFUL! PELIGRO! Since you're opening a consumer product and throwing away the case, you are also throwing away engineered safeguards. Where components were shielded with plastic, there are now exposed wires and components that could short out.**

» **Always unplug before you touch or fiddle around with anything.**
» **Heed "High Voltage" warnings. I believe the engineers! Be careful!**
» **Make sure when remounting components that they are secure and insulated.**
» **The bulb is very hot and bright. The bulb glass is a UV filter, so if it cracks, it will still power up, but you are being exposed to dangerous UV light.**

RESOURCES

Lumenlab Mega-Projector Kit:
makezine.com/go/lumenlabkit

Lumenlab basic DIY projector guide:
makezine.com/go/lumenlabguide

Lumenlab pro wiki: lumenlab.com/protectedwiki

Lumenlab forums: lumenlab.com/forums/index.php

The folks on the Lumenlab boards are very cool and knowledgeable, and have answered every possible projector question 10 times. Respect their generous sharing of information and be sure to read the builder guides and "pinned" forum topics before asking your questions. Informed, polite questions are answered pronto fast!

Special thanks to Lumenlabs and DAZZZLA!

MAKE IT

START ❯❯❯   Time: **Weekend+** Complexity: **Difficult**

1. STRIP THE LCD PANEL

1a. Remove the plastic enclosure. In your fury to rid the panel of the original consumer packaging, remain tender, like you are kissing kittens on a summer morning. I gained access with screwdrivers and putty knives, but every monitor is different. Be slow and deliberate.

NOTE: Every LCD panel is slightly different, but the basic components and structure are similar across brands.

1b. Remove the controls and RF shield. To gain access to the LCD glass, remove the whole back half of the monitor. Be careful while disconnecting the monitor controls and power button, as you will have to relocate them so that they're still functional.

1c. Remove the backlight and LCD screen. With the monitor cable disconnected, the backlight/LCD falls away in a single piece. Whenever I do a project like this, I use as much of the chassis as possible to take advantage of the original engineering. The bulk of the monitor controls will remain in the original RF shielding as a single component.

1d. Fold down the panel daughter board. Remove fasteners and fold down the monitor control board to allow the backlight to be removed.

Photography by Mister Jalopy

1e. Remove the backlight. Remove the screws and bend back the metal tabs to free the LCD from its backlight.

1f. Be amazed by the fragility of the LCD glass! Old plastic case at 11 o'clock, RF shield with monitor components at 12, stripped LCD panel in latex-gloved hands, and backlight, which will not be used, at 9.

NOTE: Disassembly was done on a grounded anti-static mat and I am wearing an anti-static wrist strap and a plaid western shirt. Only the western shirt is optional.

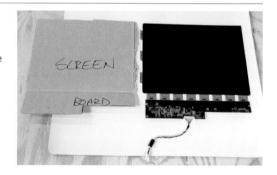

1g. Make the LCD panel dummy. To prevent damage, I made a cardboard LCD dummy to allow rough handling during mock-up of the projector. I used the natural hinge of a carton fold to replicate the flexible panel cable.

1h. Make monitor component sled. Using the original RF shielding as a component sled, I added a few pieces of wood to provide a base and another chunk of wood as a nonconductive shield between the panel and other components.

2. DESIGN THE PROJECTOR LAYOUT

Projector dimensions vary depending on the size of components selected, desired size of the projected image, and the method of box construction. A good design will allow for adjustment of internal components, as millimeters make a big difference in precision optics.

2a. Calculate the dimensions. DAZZZLA, a moderator on the Lumenlab forums, wrote a superb focal calculator application (makezine.com/go/lumenlab) to determine the dimensions of your projector based on known component values and desired projection performance.

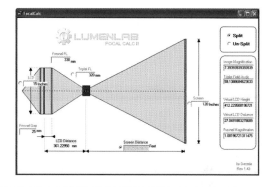

2b. Design the layout. Based on the calculator and the width of my monitor, I had a pretty clear idea of dimensions. I cut out what would become the bottom of the box and started laying out the components. Note the placement of the LCD monitor control board (aka "The Brains," as shown in bottom left illustration on page 50).

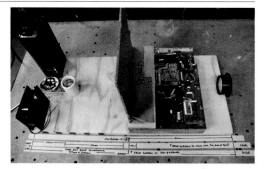

2c. Place the components. Using the calculated figures as gospel, my masking tape legend of key measurements proved invaluable for component placement.

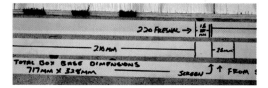

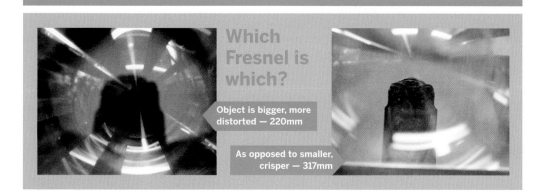

Which Fresnel is which?

Object is bigger, more distorted — 220mm

As opposed to smaller, crisper — 317mm

3. EASY SEWER PIPE LENS FOCUSING

3a. Fit the sewer pipe to the coupler. The projector triplet lens is mounted in a 40mm slice of 3" ABS pipe, which slides into a 40mm chunk of 3" ABS pipe coupler. As couplers are tapered, you'll need to wear a dust mask and sand the interior of the coupler until the pipe is snug but slideable.

3b. Screw in the lens. The wall of the 3" pipe is thick enough to drill pilot holes and screw in the lens. Use a 4" hole saw to cut the lens hole in the projector box. For a snug fit, use a strip of the fuzzy side of velcro to line the lens hole.

4. EPOXY PUTTY THE REFLECTOR MOUNT

4a. Admire the epoxy putty. Epoxy putty saves the day again! Oh, epoxy putty, is there anything you can't do? Epoxy-putty the bolt to the reflector. Under that lump of epoxy putty is the head of a standard ⅜" bolt that provides a strong, adjustable, and threaded stud to aid in the mounting.

4b. Mount the reflector hardware. A ¼" eyelet bolt is sandwiched between a pair of ⅜" nuts and washers. Though not classically beautiful, it's adjustable and it works! Perhaps I should have carved my initials in the epoxy. The finished reflector hardware is connected to a coupler nut and held fast with a pair of standard nuts. The ¼" threaded stock then continues through the bottom of the cabinet.

5. HANG THE HEAT SHIELD, FRESNELS, AND LCD

When trying to solve conundrums like how to mount the heat shield, I wander hardware store aisles with open eyes and mind. When I found a DIY window screen frame kit for under $10, I am reasonably sure that I heard angels sing.

5a. Cut the Lexan heat shield and Fresnel to size.

TIP: Score repeatedly with a sharp razor knife, then break over a table edge.

5b. Mount the heat shield, Fresnels, and LCD.
The aluminum window screen frame is the perfect channel to hold the Fresnels and the Lexan heat shield. My "brilliant" design detail of using yardsticks as cabinet material precluded me from using the window screen channel to mount my LCD screen. Makers should learn from my mistake and use the aluminum channel to hold the LCD.

FINISH ⊠

NOW GO USE IT »

So, does it work? Works great! This
is an un-Photoshopped photograph of a freeze screen from the visually stunning *The Fifth Element*. There are some caveats worth noting before you

throw away your television. A projector screen reflects light, which is great when it's reflecting your movie, but it also reflects any and all ambient light. In other words, a room that is as dark as a movie theater is almost mandatory.

Also, my projector is mounted so that it's a straight shot to the screen, but if the projector were tilted up or down, the projection would flare out at the top or bottom. This is an optical effect called keystoning, but you can eliminate it by tilting the 317mm Fresnel. Read about it on the Lumenlab forums. And read about the state-of-the-art projectors based on HDMI 10.6" LCD panels. Prepare for your jaw to drop!

» What about that nutty roof? Mister Jalopy breaks it down in CRAFT, Volume 05, hitting newsstands on Nov. 6, 2007. Stay tuned!

PROPELLER POWER
Can a wind-powered vehicle travel faster than the wind propelling it? MAKE built one to find out.

Photography by Charles Platt

The Little Cart That Couldn't

Wind-powered vehicle claims look like hot air.
By Charles Platt

Can a wind-driven vehicle outrun a tailwind? A Florida tinkerer named Jack Goodman claims it can, and he put a video on YouTube to prove it. Some sailing enthusiasts became convinced that he was on to something, while others were equally sure it was a hoax — and I decided to put it to the test.

HOW BOATS WORK

First, we need a quick primer on seafaring aerodynamics. The oldest boats had rectangular sails that were simply pushed by the wind. This system prevailed for about 4,000 years, until the 9th century A.D., when some inventive mariners found that a triangular sail enabled boats to move into a headwind at an angle, and when the sail pivoted around a mast, they could turn back and forth to tack a zigzag path directly upwind.

Figure 1 (page 62) shows a sailboat traveling north, into a wind blowing from the northwest. If the mariner adjusts his sail at a shallow angle to the breeze, the air creates a force that can be split into two components, one pushing the boat ahead, the other trying to push it sideways. If the boat has a rudder and a keel to fight the sideways force, it moves ahead. The sail's forward curve also acts as an airfoil, slowing the air in back while forcing it to move faster around the front. This adds lift to the forward force.

The idea of sailing into a partial headwind seems counterintuitive, as if we're getting something for nothing. Perhaps this explains why some boating folk seemed so willing to believe Jack Goodman's surprising claim. And because triangular sails lose their advantages in a tailwind, modern boats do better sailing across the wind than dead downwind, when you might think a boat would go the fastest.

Given how effectively boats travel upwind, and that they can even exceed wind speed while running across the wind, it seems unfair that when they go the same direction as the wind they'll never overtake it. Thus, when Goodman claimed he had a way to overcome the frustration of being overtaken by a tailwind, people were predisposed to listen.

To demonstrate his idea, he built a three-wheeled cart on which he mounted a big, homemade propeller, with its shaft connected to the rear wheels via a toothed belt. He made a video (makezine.com/go/jackgoodman) showing this contraption rolling along, first in response to a tailwind, and then, amazingly, moving faster than the tailwind, as demonstrated by a flag hanging in back. Note that if this happens, the vehicle begins to encounter a slight headwind relative to its motion.

Goodman claimed that the cart could do this because its wheels turned the propeller, which pulled it forward. This was somewhat puzzling because there was no power to turn the wheels. Was he just telling whoppers and pranking his sailing buddies? In the video, it all seemed to work, leading many people to believe that a guy in a garage in Florida had come up with something that had never occurred to anyone else in the history of boats and aviation, from the ancient Greeks through Leonardo da Vinci and on up to Northrop Grumman.

Clearly, it was time to build the MAKE version.

THE WIND TEST

Instead of a full-sized replica, I constructed a tabletop model. It would be easier to test and should perform better, since a size reduction

MATERIALS

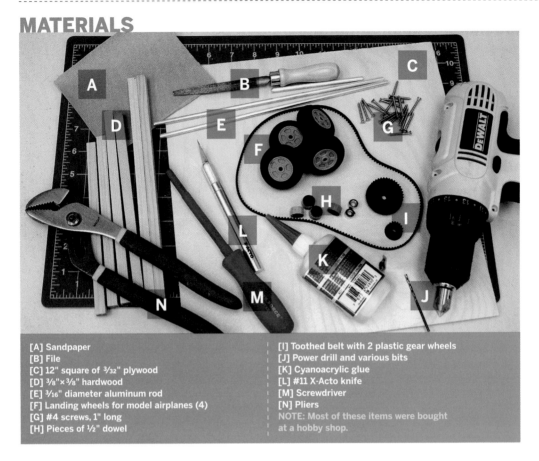

[A] Sandpaper
[B] File
[C] 12" square of ³/₃₂" plywood
[D] ³/₈"×³/₈" hardwood
[E] ³/₁₆" diameter aluminum rod
[F] Landing wheels for model airplanes (4)
[G] #4 screws, 1" long
[H] Pieces of ½" dowel

[I] Toothed belt with 2 plastic gear wheels
[J] Power drill and various bits
[K] Cyanoacrylic glue
[L] #11 X-Acto knife
[M] Screwdriver
[N] Pliers
NOTE: Most of these items were bought
at a hobby shop.

creates a lower, more favorable, ratio of mass to surface area.

I found all the necessary materials at a local hobby shop. The crucial items were a toothed belt and two plastic gear wheels, sold as spare parts for a model car. If you want to build your own cart, a plain belt will be sufficient because very little torque is involved.

To fabricate the propeller, I did what Goodman seemed to have done. I stacked a dozen wooden strips, rotating each one about 4° before adding it to the rest. This creates a very wide blade, which should work well at slow speeds. I used cyanoacrylic (aka "Krazy") glue to hold the layers together, and then smoothed them with a belt sander.

I sprayed copious amounts of WD-40 on anything that moved, then pushed the cart manually to make sure the wheels and propeller turned freely. I had to loosen the belt slightly so that it didn't create friction by pulling the shafts against their bearings. Finally, I put the cart on a workbench and subjected it to a blast of air from a 15" room fan.

The belt on the cart has a 90° twist in it. Depending on which way you make the twist, a tailwind from behind the propeller will either turn the wheels forward, so that the cart moves with the wind, or backward, so that it tries to reverse itself into the wind. With the belt twisted to enable forward motion, the cart barely moved in response to the fan. When I twisted the belt the other way, the cart didn't move at all.

Maybe I needed a bigger fan. I dragged in a 38" monster used for ventilating large warehouse spaces. Even with the new fan running at high speed, the cart ignored it and did nothing. Theorizing that the blast of air was too dispersed, I made a shroud to funnel the air and increase its speed, aiming it point-blank at the cart.

When I twisted the belt to make the cart run forward, it started moving at about 2" per second. This was not very impressive, since the air was moving at about 30' per second.

I twisted the belt the other way, and now the cart was just able to back up into the wind, which was

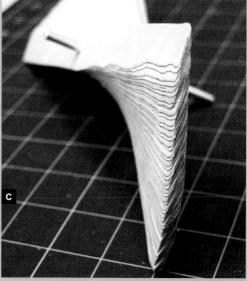

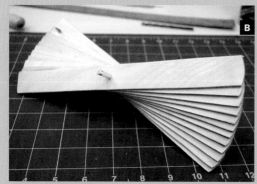

Fig. A: To make the propeller, cut 1"×10" strips of 3/32" plywood, drill a center hole, and stack the strips on the 3/16" aluminum rod, turning each one slightly before gluing it. Fig. B: The propeller completed, before sanding. Fig. C: The propeller profile after sanding. Fig. D: To join the sections of the frame, drill ample guide holes before driving in #4 screws. Fig. E: The frame completed, with plywood gussets. The uprights bend apart just enough to insert the propeller on its shaft. Holes for the shafts should be 7/32" to allow them to turn freely. Fig. F: The completed cart, ready for testing. Plastic gears are attached to the shafts with cyanoacrylic glue. The wheels are a tight push-fit. The belt must be relatively loose to minimize shaft friction.

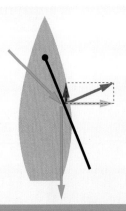

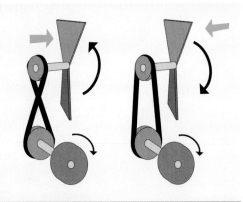

Fig. 1: SAILING INTO THE WIND
Wind (blue arrow) bounces off sail (black line) creating a force (gray arrow). The force is equivalent to two perpendicular forces (red and green arrows). The dotted rectangle shows how to calculate their strengths. The keel and the rudder of the boat oppose the lateral force, leaving the boat experiencing a net force forward. Since the wind strikes the sail over a large area, the total forward force is larger than this diagram suggests.

Fig. 2: REVERSING THE PROPELLER'S SPIN
A tailwind will turn the propeller one way, while a headwind will turn it the other way. To make the cart continue moving forward, the belt configuration must be changed at the point where the cart outruns a tailwind and encounters a net headwind.

interesting. The propeller was driving a small gear that connected by a belt to a larger gear, and this gave the wheels just enough mechanical advantage against the floor to overcome the wind pressure on the vehicle.

The cart crawled backward about an inch, and then stopped. I had been running the fan on medium, so I increased it to high, at which point it ripped the shroud off and blew it across the room. That was the end of the wind test.

AND YET IT MOVES!
Then I tried one more experiment. I attached a nylon thread to the cart and towed it along without bothering to blow air at it at all. Now the cart looked as if it was really cranking. In fact, it looked just like the YouTube video of Goodman's cart!

When I took another look at this video, I noticed that he framed it so that you never see the road more than a couple of feet in front of his cart. This made me wonder. Did he, perhaps, have a bicycle just off-camera, dragging the cart on a piece of fishing line?

What made me suspicious was not just the hopeless performance of my own model, but the belt-twisting issue. Figure 2 illustrates the problem. You can't have it both ways: if you twist the belt to make use of a tailwind, it won't work in a headwind,

and vice versa. To get the cart rolling, you have to put it in tailwind mode. Let's suppose the wind diminishes suddenly, so that the cart is now running faster than the wind. In other words, it starts to experience a net headwind. The propeller will now blow into the wind, stopping the cart in its tracks, unless you reverse the belt to take advantage of the headwind.

In the Goodman video, no one touches the belt, no gears shift when wind speed is achieved. Yet his cart just trundles along without any concern for what the wind is doing, as if — well, as if it is being moved by some other force entirely.

Perhaps Jack Goodman has some clever explanation for this. Perhaps I didn't build my version exactly the same way that he built his. Perhaps you should build your own, just to make sure. Building a prototype that doesn't work is always educational, provided, of course, that you are willing to face facts and admit that it doesn't work.

But if you decide to fake it and post a video that makes it seem to work, causing excitement and argument while spreading delusionary ideas, that is not what I would call educational.

Charles Platt is a frequent contributor to MAKE, has been a senior writer for *Wired*, and has written science fiction novels, including *The Silicon Man*.

Illustrations by Charles Platt

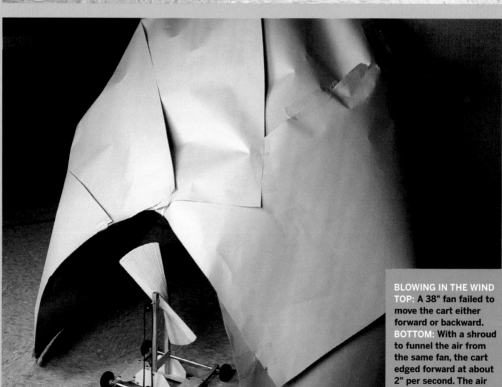

BLOWING IN THE WIND TOP: A 38" fan failed to move the cart either forward or backward. **BOTTOM:** With a shroud to funnel the air from the same fan, the cart edged forward at about 2" per second. The air was blowing at about 30' per second. Outrunning a tailwind? No way.

Swiveling Balcony Hoist

Take the sting out of a walk-up apartment by installing your own lifter. By Matthew Russell

Russell hoists his bike up to his apartment, thereby avoiding the stairwell wrestling match. The hoist is great for hauling up groceries and taking out the trash, not to mention moving in (or out).

Apartment-style living has its advantages, but getting a bicycle off the balcony, through your living area, and down several flights of stairs just to go get some exercise isn't one of them.

Carrying groceries up and taking trash down several flights of stairs isn't so much fun either, but you can mitigate these pains by building a swiveling hoist on your balcony.

The whole setup costs right around $50, and takes only a few hours from start to finish.

MATERIALS

[A] 6'–8' of 1½" metal pipe	[D] 1" metal tee	[NOT SHOWN]	TOOLS
	[E] 1½" flange	1'×1' squares of 1" wood scrap (2)	Pliers
[B] 3'–4' of 1" metal pipe (2)	[F] 1" flange	3'–5' strip of ½"×1" wood	Drill
			Screwdriver bits
	[G] Hose clamps (3)	Strong, lightweight rope	Hole saw set
[C] 1'-long, 1"-wide male-male metal pipe nipple		long enough to reach the ground floor	Carpenter's pencil
	[H] U-bolt		Hacksaw
			Tape measure
			Stepladder

1. CHOOSE THE LOCATION

Measure the distance from the floor to the ceiling of your balcony. Make a note that your hoist's inner stem (1" metal pipe) should be at least half the length of its outer stem (1.5" metal pipe). Select a mount point for the stem of the hoist. On many balconies, the area adjacent to the rail fixture is ideal because you're able to clamp to the rail and gain some additional stability.

2. MEASURE YOUR SPACE

Starting from the stem's mount point, calculate the approximate distance the arm of the hoist will need to swing out from your balcony in order to provide sufficient clearance when you're lifting up an item. If you plan to hoist up any bulky items, be sure to account for any possible rotation on the way up.

3. PREPARE THE PIPES

Thread one end of the outer stem and attach the 1½" metal flange. (Home warehouse stores usually cut and thread pipes for free if you purchase them there.) Then thread one end of the hoist's inner stem.

Connect the inner stem and the metal pipe nipple into the 1" metal tee, so that they're in line with one another. Insert the longer segment of the inner stem into the outer stem. The resulting apparatus will stand vertically on your balcony. Trim it so that it's approximately ½" less than your balcony's height. Make sure to thread and trim your hoist's outer arm.

Photography by Matthew Russell

4. DRILL THE HOLES

Use a drill and hole saw attachment to bore holes approximately ¾" deep into the 2 wood squares. Each hole should accommodate one end of the stem.

Attach the small wood blocks to the corresponding end of the stem apparatus from Step 3, and stand it up vertically. Make small adjustments to the metal pipe with the hacksaw until the full stem fits flush between the floor and ceiling. Ensure that the inner stem swivels comfortably by widening the hole in the upper wooden block as necessary.

5. INSTALL THE HOIST

Use the strip of wood and the hose clamps to secure the stem flush with the rail fixture. If no rail is available, consider an alternative means of structural support. Securely screw the wooden blocks into place. Attach the remaining 1" flange to the hoist's arm and tighten the U-bolt approximately 2" from the end. Wrap any visible threads with thin tape to prevent fraying of your rope.

6. LIFT AWAY!

Drape the rope over the arm, between the U-bolt and the end flange. Securely attach the hoist's arm to its stem. Swivel out from your balcony, and, making sure no one is below, use the rope to lower down a series of lightweight items before attempting heavier items such as a bicycle.

TIP: You can enhance the arm with a small pulley if you have the leeway.

⚠ NOTE: Safety first whenever using your hoist. Use common sense and never hoist people or animals. Test your hoist's weight limit carefully and incrementally.

Matthew Russell is a computer scientist from northern Virginia. Hacking and writing are essential to his Renaissance man regime.

Rolling Solar

Turning a junker into a sun-charged electric vehicle.
By Ben Shedd

A

B

Fig. A: John Weber's solar car gets its charge from a $350 solar panel he picked up at Costco. He bought the motor, parts, and instructions from e-volks.com.

Fig. B: To run the motor, Weber uses eight 6-volt batteries, for a total of 48 volts. Three of the batteries (not shown here) are in the front engine space.

WHILE JOHN WEBER WAS LIVING ON A sailboat in Mexico, it did not escape his notice that he got all his power from a pair of solar-charged 12-volt deep cycle batteries. When he moved back to Idaho, he decided to make a solar-powered electric car.

He bought an electric motor, drive parts, and instructions from Wilderness EV (e-volks.com). He bought a $350 solar panel online from Costco. He picked up a 260,000-miles junker from an abandoned tow lot. The first step: taking out the unneeded parts: engine, gas tank, exhaust, muffler, and radiator — as Weber puts it, "all the oil-coated garbage."

He put together eight 6-volt batteries in series to run the motor and a deep cycle 12-volt battery to run the "regular" car stuff like the turn signals and headlights, and added an electric charger for cloudy days. He had two welding jobs done on the car — four brackets on the roof to hold the solar panel and five additional battery brackets for the electric power.

The car runs quietly and smoothly. On the dash is a voltmeter instead of a gas gauge. Weber has a handwritten voltage list to determine when it's time to park the car in the sun to recover the batteries to full charge. Once it's "topped off" with energy from el sol, the car will go 10–20 miles in the city or 40–50 miles on open road.

The build took three or four months of weekend work and waiting for parts. Weber figures the whole project, including the junker car and the low-cost bright yellow paint job, cost him about $7,000 — and he's not only recycled a car, he's got zero fuel costs for it. With everything ready to go and assistance from a welder, Weber estimates he could build a second solar-powered car in a weekend — not including the paint job.

Ben Shedd is an Academy Award-winning science documentary maker. He wrote "The Year People Learned to Fly" on page 84 of this volume of MAKE.

Photography by John Weber and Ben Shedd

Granny's Nightmare Chopper Bike

Chop an old ladies' bicycle into something evil.
By Brad Graham

You know what irritates me? I drive out to the dump, pay my five bucks to get in, and the only bikes laying around are those goofy granny bikes from the late 1970s. OK — enough whining — a real chopper artist can chop any bike, even a crusty old codger cruiser.

Here's how I hacked and welded a granny bike into something evil. Normally, I wouldn't even bother with a bike like this because it has a lugged frame. This means that the head tube and bottom bracket are just press-fit and brazed into place, rather than welded. You can't salvage lugged joints for most projects because they have holes where the tubing fits together, and brazing filler metal interferes with arc welding. But this project was doable as long as I kept most of the frame intact. ✣

MATERIALS

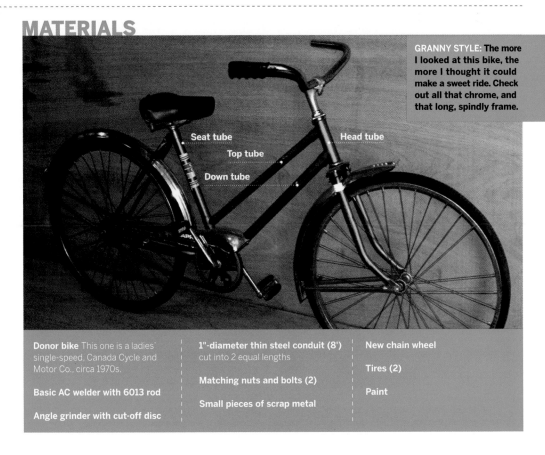

GRANNY STYLE: The more I looked at this bike, the more I thought it could make a sweet ride. Check out all that chrome, and that long, spindly frame.

Seat tube   Head tube

Top tube

Down tube

Donor bike This one is a ladies' single-speed, Canada Cycle and Motor Co., circa 1970s.

Basic AC welder with 6013 rod

Angle grinder with cut-off disc

1"-diameter thin steel conduit (8') cut into 2 equal lengths

Matching nuts and bolts (2)

Small pieces of scrap metal

New chain wheel

Tires (2)

Paint

MAKE IT.

THE FORK

First take your donor bike completely apart. Even though my donor was older than time itself, it came apart easily. The chrome parts had only slight surface rust, which I cleaned with steel wool.

I didn't want to change the bike so much that it lost all of its original look — the idea was to make it radical, yet show its roots. The main modification was turning the original fork into a long, chopper-style "triple tree" tubular fork. Since this was the focus of the conversion, I worked on the fork first and then made frame adjustments to match.

Fortunately, bikes from the good old days were made out of heavy, mild steel and were built to last, which makes them easier to weld to than the whippersnappers manufactured today from very thin steel or aluminum.

Using an angle grinder with a cut-off disc, amputate both legs off the original fork, right at

the crown (Figure A, following page). Then cut off the dropouts, keeping them even to ensure wheel alignment, and leave some extra tube above to weld them to the new fork legs. I saved the original fork legs; with those nice curves, I knew they would add some class to the frame later.

For the new fork, start with two 4'-long, 1"-diameter, thin-walled, steel conduit tubes. The tubes must be exactly the same length, and 4' is plenty. Much longer, and the frame would need radical manipulation in order to prevent the "instant wheelie" effect while riding over bumps.

I used the original fork crown as the base of the triple tree and welded the 2 pipes onto either side. To make a strong weld, grind semi-circular indentations into the crown where the tubes will fit in (Figure B).

Once both sides of the crown are ground out, lay the unit on a flat surface and tack-weld the new legs into place so that you can realign them later (Figure C).

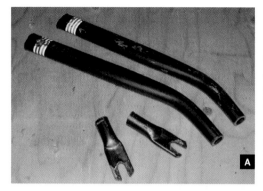

Fig. A: Legs and dropouts cut from the original fork.
Fig. B: Grinding out too much of the crown would make the fork legs too close together for a front wheel, and would invite major warping in the pipe while welding.

Fig. C: New fork legs tack-welded in place on either side of the original fork crown.
Fig. D: Welding the dropouts to the fork legs.

Next, bolt the dropouts back onto the front wheel, making sure they're perfectly parallel. If the flat parts are not in line with the tubes, straighten them with a pair of pliers.

With your new fork laying on a flat surface next to the upright front wheel, position the dropouts so they hang down and touch the ends of the fork legs. Tack-weld the dropouts in place, then check alignment by looking at the assembly from all angles, especially lengthwise. Correct any misalignments with a small hammer, and then complete all the welds, starting with the dropouts. Keep checking the alignment as you weld.

THE FRAME

With both wheels attached (Figure D), insert the new fork back into the frame and look at the geometry. Most likely, the bike will lean back and the bottom bracket will be very high, creating a "skyscraper" style chopper. Although the bike would be rideable in this configuration, I decided to mod the frame in order to lower the bottom bracket. Since I planned

to remove the top tube and convert it from a women's frame anyway, this wasn't a big deal.

Using the cutting wheel, cut the top tube from the frame at the thick part of the lugs at both ends. Cutting it flush with the head and seat tubes would make a mess, and you can use the thicker part of the lugs to your advantage later.

The down tube needs to be bent upward to keep the bike from leaning too far backward when you add those long forks. The amount is up to you. If you like a tall chopper, try bending the tube upward 10 or 20 degrees. If you want that low-and-lean look, you could bend the tube farther, but be careful of how close the pedals may come to the ground. When in doubt, lay all your parts down to get a visual plan in your mind before you make a radical adjustment.

To adjust the down tube, slice a thin, pie-shaped wedge out of the top of the tube and bend it up to fill the gap (Figure E). If you make your cuts even on both sides, then the tube will bend up easily in perfect alignment with the rest of the frame. After

Fig. E: The frame with the top tube cut off and the down tube sliced, bent upward, and welded closed.
Fig. F: The new frame — all forked up!

Fig. G: Cutting the gusset pattern. A jigsaw is better for tight curves, but this method works faster and makes good use of a worn cut-off wheel.
Fig. H: Nuts welded into the fork legs.

bending the tube up, weld the gap closed. After welding, the down tube will actually be stronger than it was originally.

I thought it would be cool to use the original bike's fork legs to make a nice, curved top tube, converting the frame from granny style to a diamond shape. The fork tubing is actually round tubing that's been squished into an oblong shape, so I un-squished the fork pieces in a vise and then welded the wide ends together end-to-end to make the new top tube. To attach it to the frame, cut off the original head tube lug and weld it to the small, curved end of the new top tube. Weld the other end high up on the seat tube.

I had a bit of one fork left over, which I used to fill in the ugly gap from the leftover lug at the bottom of the seat tube. This little curve echoed the curve on the top tube, giving the frame a stylish look (Figure F) that formed the basis for the gas-tank-style gusset I added next. I had an old cardboard box containing bicycle safety brochures, which I promptly dumped

into the recycling so I could salvage the box. I cut a cool-looking gusset out of the cardboard, and then traced the shape onto some scrap sheet steel. After cutting the shape out using a worn-out cut-off disc on the angle grinder (Figure G), I welded it onto the frame at the joint between the top and head tubes.

THE FORK CAP
The modded frame puts the bottom bracket about 6" lower than it was before, which works well. But don't sit on the bike yet, dude! Without the top of the triple tree, the forks can't take much weight.

To make the fork cap, weld two ½" nuts into the ends of each leg. Weld both nuts flush with the ends of the pipe and center them as much as you can (Figure H). Then bolt each leg to a small, oblong metal plate, which connects to the gooseneck that holds the handlebars. Cut the plates from each end of a 1½"-wide, perforated bar bracket, and grind them concave to fit against the gooseneck. Bolt the 2 plates in place, insert the gooseneck into the stem

Fig. I: Plates welded to the gooseneck. This design lets you keep the handlebars original, rather than clamping them to a single top plate somehow.
Fig. J: Seat clamp welded above the rear wheel.

Fig. K: Seat installed in its new position.
Fig. L: Dude, this isn't your Granny's bike anymore!

tube, and weld the plates to the gooseneck. To ensure alignment, start with a few tack welds and then check it all, being careful not to hit the fork threads with the welding rod. The 2 plates should align as if they are one single plate (Figure I).

THE SEAT

With the stem tube reassembled into the head tube, the frame was finished and ready to take the weight of a rider. But following the theme, I wanted new seating that came from the original granny bike, but was butchered into something evil. So I took the original seat out of its normal position and re-mounted it behind the seat tube, putting it as close to the rear wheel as possible. To accomplish this silly feat, I welded the seat clamp to the cross-brace over the rear wheel (Figure J). The nose of the seat covered up the original seat post hole (Figure K), and this looked really good — almost planned!

WHERE IS MY BIKE, YOUNG MAN?

It was time for the first test crash. I rode around the block a few times making sure everything was good, then stripped the bike right back down and painted it the original deep blue color. I buffed up the chrome parts with some steel wool, put the bike back together (Figure L), and damn it dog, it sure looked cool! To add more wackiness to the bike, I reversed the front fender. I also replaced the front chain ring with a huge chrome unit from an exercise bike. But other than that, the fork tubes, and some fresh tires, the whole deal was made from the original granny bike. It really worked out nicely.

The bike rode well, did not rattle, and felt pretty comfortable, considering where the seat was placed. The moral of this story: when you find something in the dumpster, don't think, "Hey, look at that scrap." Instead think, "Dude, this thing will make a sick chop!"

Brad Graham has been creating unique home-built bikes since the 1980s. He hosts atomiczombie.com, which showcases custom bicycles, including his 2003 World Record SkyCycle.

Swing & Wrong-Way Bikes

Trick cycles from Cyclecide.

On a Swing Bike, Laird Rickard demonstrates how both back and front wheels rotate out of the plane of the frame.

On a Wrong-Way Bike, gears between the handlebar stem and front fork make the front wheel turn in the reverse direction of the handlebars.

MAKE Projects Editor Paul Spinrad had a great time trying to ride two trick bicycles, the Swing Bike and the Wrong-Way Bike, from the bike rodeo Cyclecide (cyclecide.com). He spoke with Cyclecide's Jarico Reesce, Jay Broemmel, and Laird Rickard about how the bikes work.

Paul Spinrad: Where did the Swing Bike come from?

Jarico Reesce: Back in the 1970s, Donny and Marie Osmond actually invested in and promoted a commercial swing bike as a wacky new bicycle for kids. This was when the Schwinn Sting-Ray was popular. But too many kids fell and got hurt, and the bikes didn't sell. Like with any industry, innovative things get shelved and come back years later.

Laird Rickard: The trick to riding a Swing Bike is steering with your butt, which most people don't get when they first hop on. That's why they end up falling down.

JR: Laird here is pretty good at making the back wheel alternate between left and right. It looks cool when you get a rhythm going. Riding fast down hills, a Swing Bike really turns heads, which is one of our objectives with our bikes.

PS: And then there's the Wrong-Way Bike, which amazed me because I couldn't ride two inches on it, even if I crossed my hands.

JR: We tell people at our show that you need to be an ambidextrous dyslexic with attention-deficit disorder to ride it, and you have to look directly into the sun and ride as fast as you can.

We challenge the audience, saying we'll give $50 to anyone who can ride it. Of course no one can. In our show, it's always like, are we entertaining the audience, or are they entertaining us? Everyone gets a chuckle when some über-biker type in spandex gets on the Wrong-Way Bike and just flops and falls.

PS: Can anyone ride it?

JR: Yes, we have a clown who can ride it, Otis. And he fits the criteria of being ambidextrous, dyslexic, with attention-deficit disorder.

Jay Broemmel: I got the idea for the Wrong-Way Bike from David Apocalypse, who said it was an old carny trick. He would have the gears covered up, and charge people to try. He'd be like, "Two dollars! All you have to do is ride this bike ten feet, get across that line, and I'll give you 50 dollars!" Then he'd ride it himself and say, "Look how easy it is!"

PS: Those old carnies!

JR: Of course, we don't consider ourselves carnies. We consider ourselves showmen. If anyone calls us

📷 See more photos of Cyclecide's bikes at makezine.com/11/cyclecide.

Photography by Sam Murphy

U-G-L-Y Your Bike

To deter thieves, camouflage your bicycle as a piece of crap while keeping it a first-class ride. By Rick Polito

ABOVE: The author's piece of crap. Or is it? FACING PAGE: A Bianchi Milano without the urban camouflage.

Nature is the master of disguise. The tiger swallowtail caterpillar starts out camouflaged as a bird dropping to discourage hungry birds. Take a tip from the crawling turd and keep your bike from getting swiped: dress it down as a pile of rolling junk.

Having an ugly bike doesn't mean having a junky bike. Looks and performance have no exclusive relationship. A savvy bike thief may see the gem under the Krylon, but he also knows he can't sell it as quickly as the tricked-out speedster at the other end of the bike rack.

Photograph by Sam Polito

A. Start with the paint. A can of spray paint is a good start, but choose wisely. Black is out. We're talking ugly here, not cool. Think orange. Think brown. Think orange and brown. Use whatever you have lying around from your last project. And don't stop with the frame. The cranks, the handlebars, the tires, anything is fair game. To keep the spray-on hues from seeping into the componentry, use a Q-tip to apply a greasy line of defense.

B. The trick with paint is to make it sloppy. A coating of car wax prior to your artistic moment will turn even the most careful application into a flaking, blistering, peeling mess.

C. Bring on the rust. Rust is the enemy of bikes, but you can ally yourself with the persistent creep of oxidation by stealing from the arsenal of the faux finish enthusiast. Modern Options makes a two-bottle Rust Antiquing Set. Two minutes with this stuff is the equivalent of leaving your patio furniture in the surf line for two years. But be careful. Go for the tooth-brush splatter technique and don't overplay your hand. This isn't a faux finish, it's a fugly finish. When the rust is set, another pass with the spray can gives it that creeping-up-through-the-paint look.

D. Accessorize. Look at your seat. Now tear it. Now tape it up. Now tear it again, and pull out a bit of the stuffing. Give the handlebar grips/tape the same treatment. The components are what make the difference between a cranky-shifting clunk bucket and a smooth machine. Just make sure yours don't look smooth. Grind the logos off your derailleurs and brakes. Head to your local community bicycle co-op and pick up a pair of mismatched pedals.

E. Think specks. And just because none of these parts can actually rust doesn't mean they can't *look* like they're rusting. Be subtle here.

F. Switch out your nuts and bolts. The quick release on your seat post is a "steal me" sign. Replace it with a common bolt. For an extra level of security, switch out all the bolts you can for bolts with Torx heads. Torx is that funny Allen key/screwdriver alternative that you can never find in your toolbox. The thief left his at home. He's probably not carrying a screwdriver either.

G. Quick-release clampdown. Some urban bikers will tighten a hose clamp to hold the wheel quick releases tight to the frame. You're only buying time here, but if you don't like pulling your front wheel off every time you lock up your bike, it's an option. If you don't want to spring for dedicated theft-resistant quick releases, slip a Torx bolt into the clamp.

H. Stickers, and lots of them. With stickers, the ugly options are endless. They can make the thief think about how much time he wants to spend removing them. If you're especially sneaky, you can put Huffy stickers on your Serotta (find assorted bike decals at signwavedesigns.com). The same sticky mess factor goes for duct tape, but make it ugly duct tape. Did you know that duct tape comes in monkey-puke green? Or this might be a moment to get in touch with your My Little Pony roots and zip-tie a pink plastic basket to your handlebars.

Just hold back on any artistic effect. Don't go all Burning Man, or some poser is going to cop it for cool points. And remember, you're not theft-proofing your bike. You still need a good lock, or maybe two — many people carry a U-lock and a cable lock on the assumption that few bike thieves are carrying tools to defeat both.

But the determined thief can steal any bike, at any time. Search "bike thief" on youtube.com to see the Neistat brothers' alarming opus on how easy it is to steal your ride in broad daylight. You can't stop a thief from stealing your bike. You can only stop him from wanting your bike. And making your bike look like a piece of crap may just be the ticket.

Rick Polito is a freelance writer and dedicated bike commuter living in Boulder, Colo.

Photograph by Daniel Carter

ROAD RAVE Paul Freedman's latest Soul Cycle encloses its sound system in a glowing, molded fiberglass cabinet. The chopper frame's hinged seat post lets you switch between hill-climbing, commuting, and low-rider modes while you ride.

Photograph by Sam Murphy

Rock the Bike

Social biking with Fossil Fool and the Juice Pedaler.
By Paul Spinrad

IF YOU'VE NEVER BEEN ON A BICYCLE cruiser ride before, here's what happens. You and your party gather someplace in town and maybe have some beers and/or dinner. Then you ride. The city and the night are yours. If you're lucky, one of you is on a Soul Cycle, supplying the soundtrack.

The audio is thrillingly crisp, far better than any car stereo, thanks to no engine and less tire noise. You pedal, steer, and move to the beat, and some of your friends sing along, but others chat and laugh, steering in and out of different conversations. As the sky darkens, Down Low Glow tubes spill rolling pools of vivid, colored light onto the pavement below. Every pedestrian that your group passes smiles, waves, sways to the beat, or shouts "Right on!" and throws you a thumbs up. (But interestingly, people who are shut inside of cars tend not to acknowledge you.)

"Bikes are better social tools than cars," notes Paul "Fossil Fool" Freedman, the San Francisco-based inventor, alpha rider, and merchant of the Soul Cycle in its many incarnations. "You can stop anywhere, talk, interact. If you ask a driver whether they want more cars on the road, they'll say no, but ask a cyclist if they want more bikes, and they'll say yes."

Freedman's Soul Cycles are ultimate party bikes, low-slung cruisers with slick lighting and detailing, seating for two, and gut-thumping sound systems. Freedman's original inspiration came from a manager at the bike shop where he worked during high school, Buddy Bob, who lit up local night biking events with a recumbent that towed a stereo and a beer keg.

A few years later, after he graduated from Harvard, Freedman moved to San Francisco and started working for Xtracycle, a company that makes "sport utility bicycle" frame extensions.

Using car stereo components, the Xtracycle gang put together their Salsa Cycle for a promotional tour of Utah and Colorado, and on a warm summer's night riding along to James Brown's "Same Beat," Freedman was hooked.

Before the Soul Cycle, bike audio meant either tinny

little systems that clip onto handlebars, or clunky assemblages clamped to racks. Freedman started refining the form, putting controls in front to make them reachable while riding. The latest ones transmit the music signal wirelessly from bike to bike, to enable mob surround sound. Soul Cycle backrests, which house the amp, mixer, and battery, have evolved from wood boxes to rounded bamboo shapes to sculpted fiberglass with colored lights.

Freedman's new Chopper model dispenses with the Xtracycle base completely, substituting a custom frame he co-designed with Curtis Inglis. The Chopper also has a curved, laminated, carbon-fiber and bamboo seat post that you can swing and lock into different positions while riding, to switch between hill-climbing, commuting, and low-rider cruising mode.

Freedman's custom Soul Cycles are priced for high rollers, but for the rest of us he sells products that let you turn your own ordinary bike into a cruise-compliant party machine. The Down Low Glow is a ground-effects kit that splashes colored light down

Photograph by Paul Spinrad

onto the pavement, a rare instance where something unimpeachably cool also enhances safety.

The Soul Cycle Head Unit, based on the hacker favorite T-Amp amplifier and a Rolls Mini Mixer, hooks up to your own speakers to play music and other inputs. As Freedman explains, "I start with fantasy projects, and learn from those how to scale back for products."

Through Xtracycle's nonprofit spinoff Worldbike, Freedman met Nate "The Juice Pedaler" Byerley, another biketrepreneur with complementary experience in sustainable party technology. After moving to Berkeley in 2001, Byerley envisioned selling some sort of food off the back of a bike. He determined that tacos were unworkable, then read about a bike-powered blender in Humboldt. Intrigued, he built one himself, running the blender off of a friction wheel on the rear tire, and securing the pitcher with a wooden collar on the back platform of an Xtracycle. Holding the pitcher tight proved difficult, but Byerley discovered the perfect part to replace the wood collar: an offset closet flange, a common toilet tank component made from ABS plastic.

With the new part, the Byerley Bicycle Blender was ready to take on the road. During the summer of 2005, Byerley showed it off at ecology and music festivals throughout the West Coast and demoed it on MTV, selling numerous blenders and many more $5 human-powered smoothies. Byerley has been selling his bicycle blenders ever since.

Byerley's ingenuity isn't limited to pedal power. Soon after his wife, Kaety, gave birth to their daughter, Davis, this past January, Byerley built a baby bike seat. Kaety was skeptical until she saw what he had created: a secure platform over the front wheel with a rear-facing infant car seat that keeps Davis in constant eye contact with Byerley. The three of them now ride (carefully) together, even on nighttime cruiser rides. "Like with dolphins," Byerley explains, "the baby rides in the middle of the pack."

Freedman and Byerley joined forces in 2004 as Rock the Bike, a company dedicated to their various products and projects. They moved into a space in Tinker's Workshop in Berkeley, a facility that already housed Worldbike and the Bicycle-Friendly Berkeley Coalition (BFBC). Following the example of the workspace's community-oriented owner Nick Bertoni — a Vietnam vet, peace activist, and ex-Exploratorium exhibit builder — Rock the Bike

hires local high school kids, teaches them how to use tools, and pays them to build Down Low Glow kits, Soul Cycle Head Units, the B3 Mini (Byerley's latest bike blender), and other homegrown products in runs of 50–100. To help the bottom line, Rock the Bike also sells other people's products that they like, including Incredibell bells and Brooks leather saddles.

Last year, local musician Gabe Dominguez visited Rock the Bike to see if they could build a human-powered PA system for an all-bike concert tour he was planning with his band, Shake Your Peace. Freedman and Byerley combined their expertise to produce exactly what Dominguez wanted: a 200-watt system on the back of a bike with a heavy-duty stand for stationary mode and enough capacitor power to smooth over pedaling pauses of up to 15 seconds.

Shake Your Peace relied on the system during their 700-mile bike tour of Utah in May, and invited audience members to pedal-power each song. Audiences jumped at the chance to help, and the only mishap came when a large male volunteer wearing a pink unitard pedaled too vigorously and fried a capacitor.

Freedman and Byerley have more human-power ideas to pursue: a bigger concert PA system powered by multiple pedalers. A bike with integrated, plug-in power-generation capacity. A line of premium bicycles where all the power used to cut, weld, and finish the frame comes from pedaling. As Byerley explains, "There's something about human power that people are drawn to, but it's hard to fit into the current economic system. We're trying to stretch that in every direction and find out where the niches exist." (Meanwhile, Byerley's friend Mike Taggett is working on energy-producing exercise equipment.)

Through all their inspiration and perspiration, Rock the Bike leads regular cruiser rides around the Bay Area, and their website (rockthebike.com) lists and promotes rides in other locations. Nighttime cruiser rides are payback time for all the hard work — and not just because they show off the merch. As Freedman explains, "It takes a long time to build one of these bikes, but then you're out on a ride and you get some kids dancing on a pier or freestyling to the music, and it's totally worth it."

📷 More photos at makezine.com/11/paul_freedman

Paul Spinrad is projects editor of MAKE.

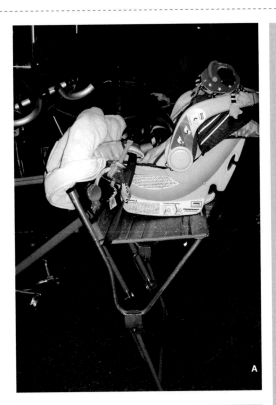

A

Worldbike: Not Just Another NGO in Toyota Land Cruisers

In 1996, Ross Evans, a mechanical engineering graduate from Stanford, traveled to Nicaragua for Boston-based Bikes Not Bombs, to teach civil war veterans there how to fix bicycles. While waiting for a shipment of bikes and parts, Ross observed how the people there used bikes to carry things. He noticed the trailers used for big loads had problems. They had extra wheels, which are expensive, and they clattered around, wasting energy while reducing mobility and handling. So Evans invented the Xtracycle, a bolt-on extension to a mountain bike that lengthens the wheelbase and adds a versatile cargo platform. This cargo bike lets you lean into turns, isn't upset by potholes, and can be ridden as an everyday bike.

Evans turned his Xtracycle into a product for the U.S. market, while his Xtracycle Access Foundation (XAF) continued to explore how it could help the developing world. In 2000, Adam French brought the XAF to Tinker's Workshop in Berkeley, and Paul Freedman took over a couple of years later, filed for nonprofit status, and changed its name to Worldbike.

In 2004, Freedman won a grant from the Lemelson Foundation, which supports inventions that benefit humanity, to test Worldbike's lowest-cost frame extension for local manufacture and use in Kenya. They called the project Big Boda, a play on boda-boda, the word for Nairobi's ubiquitous bicycle taxis.

The Big Boda project was a great experience. "People there appreciated that we weren't just another NGO in Land Cruisers," Freedman recalls. They found a great technical lead who could manufacture the bikes in Kisumu, epicenter of the boda-boda phenomenon, and did extensive test marketing in 2005 with boda-boda drivers. The Big Boda was not a universal success, but the drivers found that it worked well for loads that are bulky but not heavy, like empty water jugs and cans, and bread. Its "extended cab" also suited it well for taking children to school.

Last year, the latest version of the Worldbike was shown in the Cooper-Hewitt National Design Museum's exhibition "Design for the Other 90%."

B

C

Fig. A: Bicycle infant seat devised by Nate Byerley for his daughter, Davis. Fig. B: B3 Mini blender, with all-recycled plastic. Fig C: Worldbike market test in Kenya shows child-carrying capacity.

Bicycle
iPod Charger

A sidewall dynamo powers both lights and tuneage.
By Mark Hoekstra

After buying an iPod mini a couple of years ago, I started experimenting with ways of extending its battery life. First I tried the Perfectmate hand-cranked flashlight/charger, only to find out that it takes up to 20 minutes of cranking to generate just enough power to boot the device. As well as making me appreciate how much power today's lithium-ion batteries can hold, this got me thinking about other ways to human-power my iPod.

I LIVE IN HOLLAND, WHERE BICYCLES ARE one of the most popular means of transport. A typical Dutch bike has the old lighting system, which consists of front and rear light bulbs powered by a 6V dynamo that runs off a friction wheel on the front tire. So I got the idea to open up the flashlight charger, find out where the dynamo connects to the PCB, and connect my bike dynamo there.

The iPod mini takes USB-standard 5V and can handle up to 6V. The hand-crank charger nominally supplies 6V or a little more. Its battery pack stabilizes the current it produces, but there's no regulator to limit its output voltage. So I needed to add something to prevent the charger from possibly damaging the iPod with excess power. I first thought of a voltage regulator IC like an LM7805, but these are designed to step power back from much higher levels. Instead, I simply used a Zener diode, which

Photography by Mark Hoekstra

cuts off everything above 5.1V. I wired the output to a female USB connector to let me connect the iPod with its included USB charging cable.

For a detachable enclosure — so I could remove the system from my bike while it was parked — I considered an Altoids tin. Those are hard to find here, but I did have a couple of old Apple mice, so I decided to use a mouse case. I found that an S-video female connector fits perfectly onto an ADB (Apple Desktop Bus) connector, so I left the mouse's cable intact. Inside the case, I wired up the hand-crank charger's PCB and battery along with my voltage regulator and the USB port for the iPod. The charger also had a neat little charging light, which found its way onto the back end of the mouse.

On my bike, I connected the dynamo to an S-video plug that hangs from my handlebars. That's where I connect my mouse unit, and then I plug the iPod into the other end of the mouse. Both iPod and mouse charger are kept protected by iPod socks, one of which I embroidered with a skull design.

Then I started thinking that it would be silly to have a working dynamo on my bike but no lights, and even sillier to run LED lights with batteries while riding with the dynamo on. I did some voltage probe experimenting with an identical hand-crank flashlight, and found that when you switch the light on while cranking, it stops supplying voltage to the charging jack, and when you switch the light back off, charging resumes. This meant that my charger could easily work the same way.

I opened the charger mouse back up and connected two more wires from the ADB cable to the contacts for the flashlight charger's light. On the other side, I split these contacts into two pairs that supplied front and rear LEDs for my bike, which fit where the original bulbs had been.

Finally, I wired the mouse button to the contacts for the flashlight charger's light switch. So the mouse button has one function: switch between turning on the bike light and charging the iPod. In theory, you could run into trouble at night, with no more music and a long way from home. But during the one and a half years I've been using my charger, this has never happened.

Mark Hoekstra has a passion for technology and the urge to control and combine whatever he lays his hands on. This results in some original projects you can see at geektechnique.org.

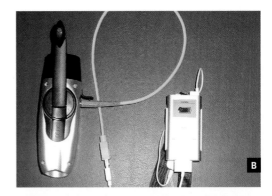

B

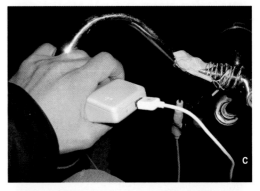

C

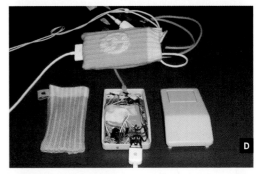

D

E

Fig. A: Bike-powered iPod charger detaches easily for parking. Fig. B: Original 6V hand-cranked charger with voltage stepped down to 5V for USB. Fig. C: Mouse case contains charger electronics and connects to bike dynamo via S-video plug on handlebars. Fig. D: iPod, opened-up charger, and iPod socks. Fig. E: Charger electronics, battery, voltage regulator, and USB port packed inside mouse case.

Stokemonkey
Makes It Easier

An electric motor linked to existing bicycle gears turns any bike into a sell-your-car-already vehicle.
By Rick Polito

TODD FAHRNER DOESN'T WANT TO MAKE IT easy for you. But he does want to make it easier.

A bike-centric, car-less existence has always been possible for strong people with strong feelings, but with Fahrner's Stokemonkey kit, it gets easier. With an electric motor linked to existing gears and an Xtracycle hitchless trailer to extend the wheelbase and cargo space, the Stokemonkey scenario turns any bike into a sell-your-car-already electric vehicle.

A committed bike commuter and car-free thinker, Fahrner knew his human-powered bikes could do anything a car could do, and better, but with parenthood making the hills that much steeper, he needed a little help. He already owned an Xtracycle-equipped bike. The hitchless trailer concept transforms commuter bikes into cargo bikes, but they aren't designed for parents living in San Francisco.

"The cargo capacity was great, but we couldn't really make full use of it in on human power alone, not in those hills with these knees as we approached 40," says Fahrner, who now co-owns Clever Cycles out of Portland, Ore.

So he took the Frankenstein approach to the best electric bike he could find, grafted on some extra gear range, and rode it into the ground for 5,000 miles. What arose from the scatter of bike parts was the first true Stokemonkey.

THE POWER OF THE TRANSMISSION
The secret of the Stokemonkey is the transmission. Many electric bikes utilize in-hub motors; some even have rollers applying power to the tire. Fahrner's design mounts an electric motor and 36-volt battery pack onto the Xtracycle-configured bike. The motor is connected by a chain drive to an extra chainwheel on the left side of the bottom bracket, boosting leg power and allowing cyclists to take advantage of a triple crankset's gear range. As a result, Stokemonkey owners can pull hundreds of pounds of passengers and cargo up the steepest hills, or zip on the flats in high gear.

That gear range gives the Stokemonkey tremendous capabilities. Bill Manewal is a homecare nurse in San Francisco. At 63, he was looking for a little oomph to maintain the no-car commute. One ride with Fahrner was all it took. "He put me on the back of his [bike] when he was living here in San Francisco and rode up this really steep hill," Manewal recalls.

Now a typical day means dozens of miles riding the hilly neighborhoods to visit patients. Speeds are upward of 25mph on flats and a good clip up steep hills, even with cargo. Manewal once delivered an industrial air cleaner to a client's home, by bike. "I ride usually between 25 and 35 miles a day, and it costs 8 cents to charge it," Manewal says. "It just levels out the hills."

The Stokemonkey kit with the motor, mount, chain, crankset, charger, throttle, and battery costs $1,350. The owner will need an Xtracycle-equipped bike. Xtracycle kits start at $244 and can be bolted onto any bike. You're not shopping for a featherweight racer here. A beater will do, but disc brakes are a good add-on. A clever parts hound could roll down the driveway for under $2,500, a lot less than even the crankiest secondhand car.

Fahrner says pre-installed, street legal Stokemonkey bikes are coming soon, but for now you have to do it yourself. He describes the required mechanical aptitude as "not much," but a Stokemonkey builder should be comfortable with basic

Photography by Todd Fahrner

TOP: **Coming home from a family camping trip, Fahrner rode a regular bike while his wife, Martina, rode this Stokemonkey bike hauling their son in a trailer, camping gear, four potted plants bought on sale, and a 55-gallon drum they picked up for a rain barrel. She matched his speed effortlessly.** BOTTOM: **Unlike other assisted bikes, the Stokemonkey puts power to the front chainwheel. It's like having an extra pedaling partner.**

bike mechanics and be confident both on the bike and at the workbench. "It has been important to me that Stokemonkey's early adopters have the strong foundational biking skills that make it safe," Fahrner says. "And I think these skills tend to go hand-in-hand with the ability to install it."

The extra power and range make the bike more capable, he says, but it's still a bike. Fahrner didn't set out to get people out of their cars. He wants to keep people who gave up their cars from getting back behind the wheel when jobs and kids tear at the convictions and the quadriceps. The Stokemonkey makes the bike capable of more tasks and utility. It doesn't make it easy. It makes it easier.

Rick Polito is a freelance writer and dedicated bike commuter in Boulder, Colo.

The Year People Learned to Fly

Clips from the 1978 Academy Award-winning documentary short, *The Flight of the Gossamer Condor*, used by permission.

Celebrating the 30th anniversary
of the flight of the *Gossamer Condor*.
By Ben Shedd

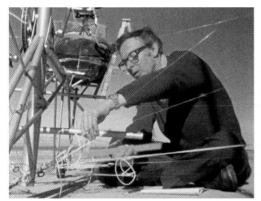

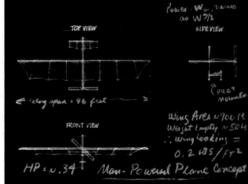

One of the great home-built family maker projects celebrated its
30th anniversary on Aug. 23, 2007: the flight of the world's first
truly successful human-powered airplane, the *Gossamer Condor*.

The plane was designed and built by Dr. Paul B. MacCready, along with his family and friends, on weekends over a year's time in 1976–77. The goal was to win the first Kremer Prize, a $100,000 reward for the first human-powered airplane that could take off using human power, fly over a 10-foot marker, make a complete left turn and right turn around two pylons spaced half a mile apart, and then fly over the 10-foot marker again at the end of the mile-long flight.

The original plans were sketches and dimensions in a notebook. MacCready, an expert in aerodynamics and a working inventor, had written an article on hang glider safety for his sons, and while daydreaming on a vacation, he suddenly realized that if he could make a hang glider with all the dimensions tripled, it would be an airplane needing only ⅓ horsepower to fly, the maximum energy that we humans can generate.

The major breakthrough was taking all the necessary large wing structure (usually a box grid for rigidity) from the inside of the wing and moving it to the outside in a huge triangulated structure — namely, aluminum poles held together with piano wire. This one idea dropped the wing loading — the weight-to-wing-area ratio — by a factor of ten over all other designs in history.

I filmed MacCready's project for my first independent documentary. I met MacCready through one of his neighbors, and when I researched other attempts at human-powered flight, I learned that if they could build it, MacCready's wing-loading formula meant this plane would do something no other human-powered plane had achieved. And build it they did — a 70-pound (empty) airplane with a 96-foot wing span.

With 120-pound bicycle racer/hang glider Bryan Allen as pilot and motor, the *Gossamer Condor* flew into aviation history, winning the Kremer Prize. The airplane is in the Smithsonian Institution's National Air and Space Museum, Paul MacCready has become known as the "father of human-powered flight," and my documentary, *The Flight of the Gossamer Condor*, received the 1978 Academy Award for Best Documentary Short.

For more information about the documentary and ordering information, visit makezine.com/go/condor.

Ben Shedd is an Academy Award-winning science documentary maker. He wrote "Rolling Solar" on page 67 of this volume of MAKE.

Maker

Spirits Guy

How Lance Winters went from basement moonshiner to celebrity vodka distiller.

By Benjamin Tice Smith

When most of us want some tequila, we run to the liquor store on the way home from work. Lance Winters prefers heading to Mexico to find the best agave cactus and bringing it back to his laboratory in a 65,000-square-foot aircraft hangar on a dormant naval base on the edge of San Francisco Bay. Winters is a craft distiller at St. George Spirits, whose vast workspace has three large stills, numerous tanks, a bottling line, and many cases of high-octane, high-priced hooch, including the top-selling Hangar One Vodka.

MASTER DISTILLER: Lance Winters sits in his aircraft hangar laboratory, in front of a shiny small-batch pot still.

It's an ideal place for making and aging liquor. The five-story ceilings, foot-thick concrete floor, and cool Pacific breezes yield stable indoor temperatures that never exceed 76 degrees Fahrenheit. But the most interesting rooms are the maze of former offices that Winters has converted into a geeky playground to entertain fellow workers and lucky visitors, as well as a refuge to tinker and experiment.

In an upstairs tasting room, bartenders, restaurateurs, and distributors can sample his latest wares. They retire to the adjoining conference room to talk business, where Winters holds forth from a repurposed ejection seat from a B-52 Stratofortress.

In another room sits a 100-year-old Chandler and Price offset printing press rescued from a Napa barn. Winters tracked down a manual and two refrigerator-sized cases of lead type for it; he'll use it to make labels for a gin he's creating, which he insists must smell like Redwood Regional Park in the East Bay hills.

That gin will be born in his personal office, where a desktop 10-liter still sits next to his computer, across from a shelf of rare books that provide recipes, wisdom, and inspiration. A leather-bound French perfumer's guide from the 1700s helps locate the most aromatic (and flavorful) portions of a fruit. Monzert's *Practical Distiller* from 1889 breaks down distilling equipment and processes. A 19th-century housewife's guide holds forgotten secrets from when making alcohol was often a DIY enterprise.

Prickly pear cactus, candy cap mushrooms, and Douglas fir are just a few of the ingredients Winters has felt were "screaming to be made" into liqueurs, even if they lacked an obvious audience outside his own adventurous taste buds. Most of these concoctions get no further than the hand-labeled bottles that clutter almost every horizontal surface. The best of these elixirs will be served in the distillery's tasting room as a unique reward to intrepid fans who cross the Bay from San Francisco. The most popular flavors make it to market, such as a brandy fortified with Lapsang souchong tea and vanilla, a vodka flavored with chipotle peppers, and another flavored with wasabi.

The cash cow that pays for all this experimentation is Hangar One Vodka. It comes "straight" as well as in four flavors: mandarin blossom, kaffir lime, raspberry, and Buddha's-hand citron. While other distilleries' flavored vodkas taste like they've been dosed with snow-cone syrup, Winters' flavors are complex and heady, to be savored like wine.

Winters goes to great lengths to preserve the smells and tastes of fresh rare fruit in alcoholic form. The stainless steel and copper Holstein pot still holds 500 liters at a time and gives off the air of a modernized German-engineered steam engine. Pressure gauges, levers, knobs, and small windows allow the distiller to control and observe the distillation process, as he tastes and smells the alcoholic steam condensing into liquor.

Behind the bubbling pipes of the biggest still sit the damp brown remains of 550 pounds of orange blossoms that have given their all to a batch of mandarin blossom vodka. Days before, the blossoms (only open ones) were handpicked and shipped in a vat of inert argon gas to prevent oxygen from sapping their essence. Now they look like a heap of yard trimmings. A few feet away, Winters opens the tank of vodka that's been steeping in the blossoms. It smells like acres of flowering citrus.

Lance Winters has a long history with potent potables. As a kid he was stymied in an attempt to mix crayon dust and water, and decided to move things along with the heat of a bare 100-watt light bulb. While most kids would have been spooked by the ensuing explosion, Lance was drawn further toward science. He spent eight years traveling the world in the Navy as a mechanic on nuclear aircraft carriers. But when he left the service he found that the skills he'd learned stoking the USS *Enterprise*'s eight atomic reactors were out of date with modern civilian power plants. He began brewing beer as a hobby.

When he edited a brewpub startup manual, in lieu of cash payment he asked for a job in the brewing industry. It wasn't glamorous, but the entry-level job at Brewpub on the Green (situated on a golf course in Fremont, Calif.) led him from waiting tables and cleaning spent hops and barley out of beer tanks to brewing and managing

ABOVE: Three German-made pot stills on the main hangar floor produce vodka and eau de vie.
BELOW: Bottles in Winters' office hold rare ingredients to add to his potions.

another brewpub in nearby Hayward.

Brewpubs stay in business by turning over a product as quickly as possible. Often it's more about squeezing a profit out of customers than the best taste out of barley. Winters wanted to make "something with more longevity, more shelf life than beer, something that could potentially outlive you." So he bought a 25-gallon pot still and began dabbling in "moonshining."

His DIY whiskey wasn't for sale, it was just another outlet for his creativity and something to share with friends. But it was still illegal. While homebrewing and winemaking in small amounts are legal, home distilling is heavily regulated. There is a lot of tax revenue at stake: the Feds make $13.50 per 100-proof gallon of spirits. States take a hefty cut as well, and they have no desire to lose that kind of revenue.

Other distilleries' flavored vodkas taste like they've been dosed with snow-cone syrup. Winters' flavors are complex and heady, to be savored like wine.

Seeking advice, in 1996 Winters brought his homemade 'shine to Jorg Rupf, an old-world distiller who had come to California 25 years earlier in search of the best produce to make eau de vie, a clear spirit that preserves the flavors of fresh fruit. Rupf was impressed by the self-taught American and his whiskey. So when Winters asked for a job, Rupf offered him a month tryout. Winters quit the brewpub for good the next day.

Of course Winters wanted to develop his homemade whiskey into an item they could sell. But before he could try, he had to learn the far more difficult process of making eau de vie. "Jorg was sort of like Mr. Miyagi to my Karate Kid. He smacked me around until I didn't know anything anymore and taught me how to make

eau de vie, which became a better foundation to make whiskey."

Eau de vie is French for "water of life" (*uisge beathe*, the Scottish Gaelic root of *whiskey*, shares that meaning). Rupf insists it be made of pure fruit, so a single bottle requires 30 pounds of raspberries, pears, or cherries. The produce must be perfectly ripe to provide flavor and sugar, but a single bruised fruit can allow nasty wild yeast strains to ruin a whole batch. Since eau de vie does not rely on aging or wood for smoothness, it has to be fermented and distilled carefully so it doesn't taste harsh coming from the still.

Distilling basically involves heating up an alcoholic brew and separating the ethanol alcohol (which evaporates first) from the water. But the trick is to control where the other elements in the mix, called congeners, go. Ideally, the good congeners — which give a liquor flavor, mouth feel, and body — rise with the alcohol, while leaving behind the bad congeners that cause nasty tastes and hangovers. Turn the wrong handle on the still at the wrong time and smooth, flavorful nirvana turns into harsh, headache-inducing swill.

Whiskey distillers leave more congeners in and rely on decades of aging in oak barrels to smooth out the rough edges. Vodka, however, is so highly distilled that what comes out of the still is 95% alcohol, leaving almost all congeners behind — throwing out the good with the bad. Then water is added to reach 80 proof. Making eau de vie is especially tricky as it must retain enough good congeners to keep the delicate fruit flavors, but remove the bad to stay smooth enough to drink without the benefit of aging.

Rupf taught Winters to make whiskey coming out of the still smooth, with overtones and complexity, skipping a decade of aging in the process. After two years of experimentation and three years of aging, they started selling a whiskey called St. George. In a few years they were selling as much whiskey as eau de vie.

Intrigued by the success of premium vodkas like Grey Goose, which sells for $30 a bottle, Rupf and Ansley Coale (co-founder of Mendocino County's Germain-Robin brandy distillery) pooled resources to start Hangar One. They started with a straight vodka that mixes in a

ABOVE: Extracts from exotic plants and fungi are used to flavor the varieties of vodka.

distillate of Viognier grapes to smooth out the base, which is distilled from wheat. With their fruit experience, Winters and Rupf knew they could do great things with flavored vodka, and their shared passion for Asian food inspired the kaffir lime and Buddha's-hand flavors. Without aging to worry about, it took months instead of years to perfect. Hangar One went on sale in 2002. Immediately, three magazines named it "vodka of the year" and now Hangar One outsells everything else they make by a margin of 40 to 1.

Winters is always searching for a new taste, and each new liquor is a complicated project, entailing far more than throwing some new ingredients into the fermenting tank. He goes on expeditions to find the best ingredients and devises new techniques to extract the flavor. He tests different recipes to discover the best way to keep the flavor intact through distillation.

Lately he's been experimenting with rum. This involved tracking down a small mill to squeeze the sweet juice out of the fibrous sugar cane. After trolling eBay for a used mill, he found a manufacturer of new mills in India. Once his new mill arrived, a power supply had to be fabricated from motorcycle parts, and a stainless steel holding tank made. He fed 40,000 pounds

of "elephant" sugar cane, grown by Laotian immigrants in Fresno, Calif. into the mill. He fermented the extracted sweet juice using Sauvignon Blanc yeast, which has grassy notes that suit rum well.

The resulting 600 gallons of rum didn't have the taste he was looking for, so he ran through another 20 tons of cane from Brawley, a hotter area near San Diego that made a sweeter, more complex flavor. The mountain of spent cane fibers (called bagasse) sits smoldering in a neighboring nursery's compost heap. But if the rum catches on like the vodka, he plans to send future loads of the bagasse to a neighboring business that will use it to make disposable cutlery.

Winters appreciates the irony that after he quit running reactors on aircraft carriers, he's ended up working on the water's edge in a Navy hangar.

"During my time stationed on the USS *Enterprise* I'd sit literally trapped on the ship and marvel at the views of San Francisco, thinking that I'd really love to have a cocktail to sip while I enjoyed the view." He laughs. "Mission accomplished."

Benjamin Tice Smith, a magazine photo editor, likes to build furniture, wire up electronics, and keep old cars and motorcycles running. He lives in Oakland, Calif., with his wife and two children.

Personal Fab

A CNC MILLING MACHINE FOR LESS THAN A GRAND

By Tom Owad

THE OBJECTS WE BUY TODAY ARE MADE in enormous factories, typically thousands of miles away. Like the mainframes that preceded personal computers, these factories are filled with expensive machinery, controlled and maintained by dozens of operators and technicians. Also like the mainframes, they're about to be displaced by home machines.

Neil Gershenfeld of MIT's Center for Bits and Atoms has written *Fab*, a book that describes the possibilities that home fabrication machines will create, and how people today are using fabrication labs that the Center created. A critical component of personal fabrication is CNC — computer numerical control. A CNC machine is controlled by a computer that issues commands for each movement the machine makes.

One of the machines most commonly adapted for CNC control is the milling machine, which resembles a drill press with a movable table. The table has x and y axes that are controlled by motors. Using a language called G-code, the programmer can direct these motors to move to a particular location, traveling on a defined path at a defined speed. The z-axis is the spindle (or the table) and can be raised or lowered to adjust the depth of the cut. Fourth and fifth axes can be added to rotate the work.

A commercial CNC milling machine will cost from $10,000 to well over $100,000 — but retrofitting a manual mill will give you a home CNC mill for a fraction of that cost.

The Sieg X2 is a very popular mini mill for CNC conversion. Made by Sieg and marketed by Grizzly, Harbor Freight, and others, it sells for about $500. The simplest CNC conversion you can do with it is to automate the x and y axes, leaving you to manually adjust the spindle to the desired level. In this configuration, the x and y axes are controlled by stepper motors that are mounted in place of the handles. You can make the motor mounts using your mill or you can buy a mounting kit, such as the one sold by littlemachineshop.com (part #2668). This provides all the brackets and couplings you need to mount your motors. It's still up to you to

provide the motors and the circuitry to control them.

There are many boards, called stepper drives, available to control stepper motors. My favorite is the Microstep by EAS (embeddedtronics.com). EAS sells the bare circuit boards and the PIC chips, and provides parts lists for ordering the rest from Digi-Key. They even provide the source for the PIC and layout for the printed circuit board, if you'd rather do everything yourself.

An easier and more popular option is the Gecko Drive (geckodrive.com). This comes fully assembled and supports more powerful motors than the EAS Microstep. Both options require an unregulated linear power supply. These are about as simple as a power supply gets, and are easy to build.

The stepper drives can be controlled by an ordinary onboard PC parallel port, but if you use a PCI parallel card, you reduce the risk of damaging your motherboard. EAS and many others sell parallel port breakout boards that make it easy to interface the stepper drivers. To control the machine, there's Enhanced Machine Controller or EMC (linuxcnc.org), a very powerful open source controller that can convert G-code into motion.

You now have an entire CNC mill, if a simple one. The costs pencil out as follows:

Mini mill	$500
Mill accessories	$100
Stepper motors (2)	$100
Microstep drives (2)	$100
Power supply	$25
Mounting kit (optional)	$265

So, about $825 and a lot of dedication will get you a basic home CNC milling machine that can cut metal, plastic, wood, and circuit boards. A fun first project might be to build your own wooden clock: pathcom.com/~u1068740.

Tom Owad is the owner of Schnitz Technology, a Macintosh consultancy in York, Pa. He spends his days tinkering and learning, and is the owner and webmaster of applefritter.com.

Make: Projects

Race cars, tiki masks, and gorgeous bird pictures, ahoy! First, relive the bugs-in-your-teeth days of vintage auto racing with this retro radio-controlled roadster. Next, learn that life in a vacuum doesn't suck when you start vacu-forming 3D parts in your own kitchen. (Masks and jello molds are just the beginning.) After that, use this remote control rotating bird feeder to help you snap the best photos of your feathered friends.

Retro R/C Racer

94

Kitchen Floor Vacuum Former

106

Rotating Bird Feeder

116

RETRO R/C RACER

By Frank E. Yost

RIVETING TALE

Using scrap sheet metal and pop rivets, you can construct a model 1930s British Midget racer that combines vintage "tether car" styling with modern R/C capabilities.

Radio control (R/C) toys are fun, but their plastic bodies are so obviously mass-produced. To create something more interesting, I wanted to find a metal toy racer 10 inches or longer that I could transplant some R/C insides into. I soon discovered that the best candidates were all precious collectibles.

I decided to build a metal body myself and soon stumbled upon an old hobby called "tether cars," a type of model racer that predates R/C (see sidebar, page 105). After ordering blueprints from a tether car enthusiast group in England, I learned how the old toys were made, and then ported the style to my R/C project. With a few modern adaptations, such as pop rivets instead of solder, and coil spring shocks, I built a couple of vintage-style R/C cars that would make any antique dealer jealous. Here's my latest.

Set up: p.98   Make it: p.99   Use it: p.105

Frank E. Yost is an amateur artist who lives in Andover, Minn. His interests include drawing, woodcarving, bronze casting, welding garden art, and building tether cars. In 2002 he won an award for his comic book *Cookie and Butch*.

Photograph by Sam Murphy

BLOCK, SHOCK, AND BUCKLE

R/C hobby manufacturers produce countless prefab parts. Some are great for combining with parts you make yourself.

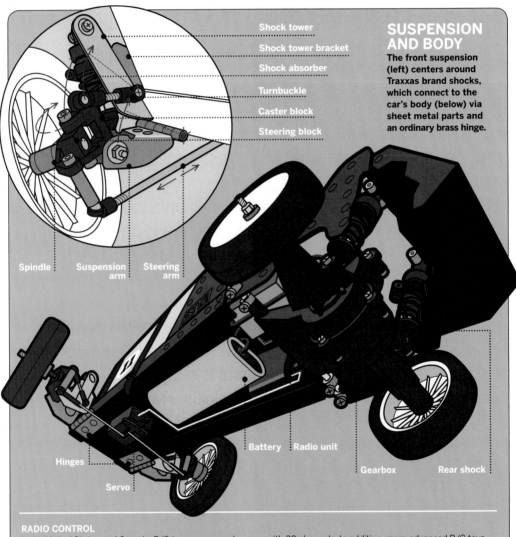

Shock tower
Shock tower bracket
Shock absorber
Turnbuckle
Caster block
Steering block

SUSPENSION AND BODY

The front suspension (left) centers around Traxxas brand shocks, which connect to the car's body (below) via sheet metal parts and an ordinary brass hinge.

Spindle | Suspension arm | Steering arm

Battery | Radio unit

Hinges

Servo

Gearbox | Rear shock

RADIO CONTROL

In the United States and Canada, R/C toys use several radio bands, most commonly the 27MHz band, which has 6 channels spread between 26.995 and 27.255MHz. When users play near each other, they avoid accidental signal hijackings by flying colored flags from their toys' antennas to indicate the channel they're using. R/C aircraft can also use a 72MHz band that has 50 channels, and surface vehicles such as cars and boats can use a 75MHz band

with 30 channels. In addition, more advanced R/C toys requiring an FCC amateur license operate at the 50MHz band (10 channels) and 53MHz band (8 channels).

R/C toys transmit control information over these carrier frequencies in various ways, including AM, FM, and PCM encoding schemes. With cars, the 2 values typically transmitted are speed and direction, which control the car's main motor and steering servomotor, respectively.

Illustrations by Nik Schulz

SHEET METAL MODELING: A MINI-PRIMER

Sheet metal is a major component of junk everywhere, and with pop rivets and simple skills, you can turn this durable material into a model racer — or almost anything else.

TYPES OF SHEET METAL

Regular sheet metal made from steel, is very forgiving: bend it wrong, and you can just hammer it flat again and start over. Thickness varies; I use between 18 and 24 gauge. Most sheet metal is coated to prevent rust. Tin-plate steel, also called simply "tin," has a uniform shine. Galvanized steel, coated with zinc, sports a distinctive crystallized pattern.

Stainless steel tends to be harder than regular steel, so it's more difficult to work. If your tinsnips aren't strong enough to get through it, try cutting with a rotary tool (Dremel).

Aluminum is lightweight and soft, easy to work, but easy to dent and scratch. Aluminum also weakens when bent in the same spot too many times. But it has a beauty unto itself, and was once even considered a precious metal.

OPERATIONS

 Bending For most bends, you should use a sheet metal brake. Cheaper brakes can let the metal slip a bit, which puts the fold in the wrong place, but you can discourage this by first applying masking tape along the fold line and clamping the metal down. To bend small pieces of stiffer metal such as stainless, clamp them in a vise between wood blocks and hammer them over.

 Cutting Make paper templates for all the shapes you're cutting. Then either stick them directly onto the metal with double-sided tape, or glue them to cereal-box cardboard, cut out, and trace around them on the metal with a fine-point marker. To help anchor larger templates, cut holes in the middle and tape over them. Cut shapes out with tinsnips or a Dremel cutting wheel, but note that a Dremel can sometimes make tape melt and slide around. File or sand the edges smooth after cutting.

 Drilling Center-punch holes before you drill, and use a high-quality drill bit with a sharp tip. I've bought some surprisingly bad drill bits, which wind up dancing around on the metal and biting in at the wrong place. Drill at ⅛" for standard pop rivets.

 Riveting Insert a rivet through your hole. For extra strength, back it with a washer on the other side. Then squeeze with the pop rivet tool. To remove a rivet, drill it out using the same ⅛" bit you made the hole with. To test-fit a rivet, you can hold it in place temporarily with a nail.

PROTOTYPING WITH CARDBOARD

Anything you want to make with sheet metal you can prototype first with cardboard. Just cut the pieces out, then fold and tape them together. This is an easy, fast, and cheap way of checking out a design before building.

 SAFETY: Always wear eye protection and tie up your hair while working with any power tools. Wear gloves during cutting; sheet metal can be sharp.

SET UP.

MATERIALS

Steel sheet metal, tin or galvanized, about 20 gauge **I used the case from an old Compaq tower.**

Thin steel sheet metal, about 26 gauge, 2'×2'

Thin aluminum **I used an 18"×13" cookie sheet.**

Thick aluminum **I used an old street sign.**

Stainless steel **I used the reinforcing brackets from an old desk, around 18 gauge.**

Perforated sheet metal **I used the ventilation panel from a very old computer.**

¹⁄₈" metal rod, about 12" long **I salvaged this from an old typewriter.**

Wire coat hanger

Old inner tube

Double-sided, masking, and electrical tape

Scrap cardboard **for templates**

Scrap wood or plywood **not balsa**

Thin wire or twist-ties

Mini toggle switch and wire **(optional)**

Double-sided foam tape **(optional)**

½"×¾" aluminum angle ¹⁄₁₆" thick, 3' length

½"×½" aluminum angle ¹⁄₁₆" thick, 4' length

¹⁄₈" aluminum pop rivets and pop rivet washers (100)

1½"×1¼" brass hinges (2)

⁵⁄₃₂" round copper tubing, 6"

Assorted small machine screws and bolts

FROM A HOBBY STORE OR R/C ENTHUSIAST

Any donor R/C car will work. **I used mostly parts from a Bandit, from a box of used parts I bought from a neighbor for $25. If your parts differ from the following, you may have to adjust dimensions and construction details to fit.**

Traxxas PARTS: Steering blocks and wheel spindles (2) **part #1837** Pro-series caster blocks, 25° (2) **part #1932**

Big bore shocks, short (2) **part #2658**

Wheels, front (4) **part #2475**

Tires, 2.1" spiked, front (4) **part #1771**

Suspension pin set, hard chrome **part #1939**

Tom Cat/Spirit gear-box assembly various parts **find one on eBay, assembled**

Hot Bodies purple threaded touring shocks (2) **part #HB24010**

Tamiya TT-01 turnbuckles **for R/C cars (2)**

2-channel radio control (R/C) system and compatible electronic speed control (ESC) **I used a Futaba T2PH with a Tazer 15T ESC, bought on eBay for $20.**

¾" turnbuckles for R/C cars (2)

Pack of miscellaneous brass bushings

Pack of miscellaneous lock collars **aka brass wheel collars or Dura-Collars**

Lacquer paint and primer

TOOLS

Pop rivet gun, sheet metal bending brake, tinsnips, rotary tool with cutting and grinding wheels such as a Dremel, hacksaw, disk and belt sander, drill and ¹⁄₈" drill bit, metal files and sandpaper, hammer, pliers (regular and needlenose), wire cutters, center punch, ruler, C-clamps, scissors, drafting triangle and tri-square, screwdriver for R/C parts, vise, extra fine-point marker, white glue, a penny, a couple of nails, protective eyewear, fine paintbrushes, soldering equipment (optional) if installing switch

BUILD YOUR VINTAGE R/C CAR

START ⋙   Time: **A Month of Evenings** Complexity: **Medium**

1. PRINT OUT THE BLUEPRINTS

Download the project blueprints at makezine.com/11/retroracer and print out at full size. The images are oversized (16"×24"), so with a letter-size printer you'll need to print multi-page and tape the pieces together.

2. MAKE THE FRAME

2a. Cut section 1A out of the blue-prints, glue to some scrap plywood, and let dry overnight. This is what we'll be building on.

2b. Use a hacksaw to cut two 9" lengths of the ½"×¾" aluminum angle. Use masking tape to hold the 2 pieces in their places on top of Section 1A.

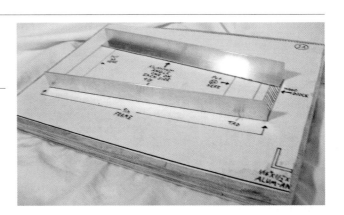

2c. Cut 3A out of the blueprints, trace the template onto tin or galvanized sheet metal, and carefully cut out the part using tin-snips. Shape by folding along the dotted lines using a brake and pliers. Trace, cut, and shape part 2A. Note that the longest bend goes the opposite way from the other 3. Sand a small piece of wood to ¹¹⁄₁₆" thickness and cut to the shape of template 4A. Slide wood block into 2A.

2d. Place the 2A/4A combo into the front of the frame and 3A into the back. Drill ⅛" holes in parts 2A and 3A where indicated by the + symbols, and pop-rivet the frame together. Trace around a penny at the front end of each aluminum piece, then use a belt sander to round out the front of the frame.

2e. Cut and drill brass hinges following pattern B7. The sides of the hinges to cut are the ones with 2 loops around the pin, not the ones with 3. Slide the uncut sides of the hinges under the sides of the frame about ½" from the front. Tape and clamp everything back down without blocking the holes on 2A and 3A, then re-drill those holes, going through the frame and the hinges. Unclamp everything, and then pop-rivet the hinges to the frame, using washers underneath for added strength.

3. MAKE THE FRONT SUSPENSION

This is the most difficult part of the project — everything must be adjusted just right. Old tether cars had leaf springs, but I wanted a buggy-style independent spring suspension.

3a. Cut out sheet metal parts B3 and B4, the suspension arms that connect the hinges to the shocks and caster blocks. Drill ⅛" holes where indicated, and shape using pliers. Pop-rivet them to the hinges. Cut the front shock towers (B1 and B2) out of something strong. These parts take a lot of stress, so don't use thin steel or aluminum. I used stainless, cutting the parts with a rotary tool and then filing them down. Use a vise and a hammer to bend the shock towers following the patterns. Bend them to only about 80° and make sure the 2 angles are the same. These towers connect the frame to the tops of the shock absorbers.

3b. Drill towers B1 and B2 with the ⅛" holes indicated by arrows on the blueprint. Pop-rivet to the car frame, over the hinges. Gauging with a drafting triangle, pivot each tower to lean slightly forward so that its top edge lies ⅜" from vertical. Clamp them in position, then drill and rivet the 2 holes next to the existing rivet to secure the towers to the frame.

3c. Cut and bend shock tower brackets B5 and B6 out of the same metal as B1 and B2. Clamp them into place on the front of the frame so that the flat sides sit parallel to the back of 2A and the side tabs sit flush against the tower. Drill and rivet the bracket tabs to the towers and the frame. Don't drill all the way through the wood; just go far enough to accommodate the rivet, which serves as a nail here.

3d. Use machine screws and bolts to attach each front shock between the top of its tower and the front of its suspension arm. Bolt the bottom of each caster block into its arm. Assemble the car's turnbuckles, substituting shorter ¾" rods for the originals, then connect one end to the top of the caster block and bolt the other to the back of the shock tower at about the same horizontal level. Connect the steering blocks to the caster blocks using suspension pins.

4. MAKE THE BODY

4a. If you're using a Traxxas Tom Cat/Spirit gearbox, trace the cab pattern 1C onto thin sheet metal, cut out using tinsnips, and flatten using a hammer and scrap plywood. With a smaller gearbox you can use pattern 2C.

4b. Smooth the edges of the metal using a file and sandpaper, then bend the cab following the pattern. Use a brake for the large bends and pliers for the smaller ones. Make sure the body is square. The blueprints also include an optional side cab, part 3C, which you can add on for looks.

4c. Trace pattern D1, for the hood, onto aluminum, then cut out slowly using a cutting wheel (don't use tinsnips). File edges. Use a brake and pliers to bend the hood along the lines indicated on D1.

NOTE: Beginners might cut piece D1 out of tin, which is more forgiving if you need to tweak the bends.

4d. Cut out template G1, the radiator grille, and slide it into the front of the hood. Trim the template to fit, then trace onto perforated sheet metal. Cut out the sheet metal grille and check the fit carefully before riveting. Drill and rivet the grille onto the hood.

4e. Cut paper template D1 into parts D2 and D3, as indicated. Tape the scuttle template, D2, onto the hood, and then trace onto the metal. Use a cutting wheel to carefully cut the hood along the line you just traced, then file down any marks.

4f. Slide the cab, scuttle, and hood into the frame. C-clamp the scuttle to the frame, remove the hood, and then drill and pop-rivet the cab and scuttle into place. Cut, smooth, and bend hood lip F1. Fit under the scuttle, then clamp it down, drill, and rivet into place.

NOTE: Be sure that F1 points slightly downward so that the scuttle will be at the same angle as the hood.

5. MAKE THE REAR SHOCK TOWERS AND TRUNK

5a. Cut two 2" lengths of the ½"×½" aluminum angle. Drill according to patterns 4C and 5C, then rivet side-by-side to the inside back of the cab, centered 1½" apart and vertical. These are the rear shock towers.

5b. Cut 2 more pieces of the ½" aluminum to 3½" in length, and drill as parts E3 and its adjacent E2, the trunk brackets. Rivet to either side of the back of the cab, so they stick out vertically with an outside distance of 3½" apart.

5c. Out of thin steel, cut out 2 trunk side pieces following template E1 and 1 trunk cover piece, E2. Drill, bend using a brake, and rivet together. Fit the assembled trunk over the trunk brackets, drill through from the sides, and rivet into place.

NOTE: We'll get to installing the gearbox and rear wheels later.

5d. For the antenna bracket, fold a 1¾"×¾" piece of sheet metal around some $5/32$" copper tubing using some pliers. Cut 1" of the tubing to remain in the bracket, then center-punch and drill holes on each side of the folded tab. Drill and pop-rivet the bracket next to the trunk. The plastic radio antenna will slide perfectly into the tubing.

6. FINISH THE FRONT END

6a. Trace windshield template H1 onto thick aluminum and cut out with a cutting wheel. Sand and file all the edges smooth. File the inside edges by hand. Trace and cut windshield bracket H2 onto thin steel. Bend the tabs on H2 using pliers, and check the bends for proper fit by placing it on top of the scuttle. Cut 2 side curtains H3 out of thicker steel, bend in a vise, cut the inside using a cutting wheel, and file smooth outside and inside.

6b. Fit bracket H2 onto the scuttle using small clamps, then clamp windshield H1 onto the bracket. Unclamp the pieces from the scuttle, then drill and rivet together. Clamp, drill, and rivet the side curtains H3 to the windshield, then position this assembly back on the scuttle and drill through the bracket and scuttle. Bolt into place using small nuts and bolts.

NOTE: Use an old rag in the vise to prevent the metal from getting marred.

6c. Optional: I wanted a switch on the dash, so I made a bracket out of stainless, following template I1, and used a large bit to drill the hole for the toggle. I bolted the switch to the scuttle, and hooked it up by cutting the receiver unit's power wires and soldering them to the switch's terminals.

6d. As in a vintage car, the hood is held down with a strap. Make it by cutting an 11"×¾" strip from an old bicycle inner tube. Then cover each end with electrical tape and cut a slit for attaching it to the hooks under the car. On another model car I used leather instead of rubber, making it more authentic.

6e. Make the hood strap hooks out of a wire coat hanger. Cut 2" pieces, then use needlenose pliers to make hooks on one side and loops on the other. Pop-rivet the loops under the frame.

7. INSTALL THE GEARBOX, STEERING, AND RADIO UNITS

7a. Cut off the ball joint pivot that protrudes from the front of the gearbox assembly. Thread a ⅛" metal rod through the horizontal sleeve on the gearbox that sits just under the old ball joint.

7b. Attach the rear shocks to the back of the gearbox, then drill and bolt the other ends to the rear shock towers. Connect the front of the gearbox by drilling holes in the frame for the metal rod to thread through, then trim the excess length and secure it with lock collars at each end. Put brass bushings onto the steering shafts, then install the front wheels and tires.

7c. Cut frame bracket 5A out of thick aluminum and rivet it flat into the frame. Bolt the servo unit to 3A and 5A so that the steering rods run underneath the frame. To boost undercarriage clearance, I raised the steering unit up on ½"-thick wooden spacers. Install the receiver and speed control boxes behind the servo, on opposite sides of the frame, under the scuttle. I cut and riveted small brackets to hold them, but you can also just use double-sided foam tape.

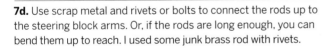

7d. Use scrap metal and rivets or bolts to connect the rods up to the steering block arms. Or, if the rods are long enough, you can bend them up to reach. I used some junk brass rod with rivets.

7e. Connect the battery unit and install it in front of the servo. I drilled holes in 3A and 5A, looped wire tightly between the 2 on either side, and used more wire to suspend the battery between the 2 loops.

8. TEST, PAINT, AND REASSEMBLE

8a. Adjust the turnbuckles so that the front wheels are straight. Confirm that everything is connected and test the car out with the radio.

8b. If everything works properly, disassemble the body pieces and paint with lacquer, starting with a coat of primer. I used Plasti-Kote Burgundy Metal Flake for most of the body and black for the inside. In honor of this issue of MAKE, I painted on a number 11.

8c. Let lacquer dry, and reassemble the car. You're done!

FINISH ☒

NOW GO USE IT ≫

MAINTENANCE AND ADJUSTMENTS

Keep all moving parts working freely, with no rubbing of any kind. If a part is rubbing the wrong way, adjust it with pliers or file it down. Also, make sure the front shocks don't press down too hard on the hinges, or they won't rebound fast enough.

Make sure the nuts and bolts that hold the caster blocks in the front suspension arms aren't too tight. The block should be able to pivot; if not, add some washers and oil to the joint.

Photograph by Sam Murphy

REMOTE POSSIBILITIES

Photograph from *Model Craftsman* magazine, 1940-1941, courtesy of tethercar.net

TETHER CARS: A SHORT HISTORY

Tether cars, also known as spindizzies, are the miniature racing car hobby predecessor to the R/C era. Cars race one at a time around a circular track, tethered to a steel pipe that sticks up in the center, and accelerate with each lap. The hobby started in California in the 1930s, and back then the 10cc gas-powered cars would often exceed 50mph. Many cars looked like the midget racers (sprint cars) that were popular at the time.

Tether cars have no brakes. To stop them, the car's owner lowers a stick, flag, or broom over the car's path as it passes underneath. This knocks down what looks like an antenna sticking up in back, but is actually a cut-off switch for the fuel or electrical supply.

During World War II, tether car racing was put on hold in the United States, but it began flourishing in England. The British government grounded toy gas-powered airplanes and kites, fearing that spotters might mistake them for Nazi aircraft (I also wonder if they wanted to hide their near-ground radar capabilities). Then in 1942, D.A. Russell and A. Galeota published a design in the magazine *Aeromodeller* for a simple spindizzy powered by a 2.55cc gas engine that was originally built for model airplanes. It was an instant success, and tether car clubs soon formed all over England.

Tether cars still have a following today, with enthusiasts divided into two camps: collectors and vintage-style builders who love the aesthetics of model making, and competitive racers more interested in competition and speed records. Traditional tether cars look like real cars, but the ones that compete today at dedicated tracks in the U.S., Europe, and Australia have more abstract designs, optimized for speed. Some of these rockets-on-wheels can top 200mph. That's fast! (But personally, I'm more interested in the artistic side.)

Plans for vintage tether cars are available in books and online (see Resources), but they're tough to build today because some of the transmission and drivetrain parts haven't been manufactured

for more than 60 years. You can still construct the bodies, though, and I think that steering and accelerating via radio control is more interesting than running around in a circle anyway. A vintage-style tether car body with modern R/C combines the best of both worlds.

RESOURCES

Spindizzies: Gas-Powered Model Racers by Eric Zausner, EZ Spindizzy Collection, 1998. Eric also runs the Spindizzy Museum in Berkeley, Calif. ezspindizzies.com

Vintage Miniature Racing Cars by Robert Ames, Graphic Arts Center Publishing Company, 1992

Reprints of old British and U.S. tether car books: tethercar.com

American Miniature Racing Car Association: amrca.com

Retro Racing Club, Louth, Lincolnshire, U.K.: +44 (01507) 450325

More resources at makezine.com/11/retroracer.

KITCHEN FLOOR VACUUM FORMER

By Bob Knetzger

FORMING LASTING IMPRESSIONS

From take-out coffee lids to airplane interior panels, vacuum-formed plastic is everywhere. And for good reason: vacuum forming makes light, durable, and cool-looking 3D parts. Here's how to cook some up in your kitchen.

My favorite childhood toy was the Mattel Vac-U-Form. The pungent smell of melting plastic filled my bedroom as I spent many hours molding little cars, bugs, and signs. The way the flat plastic changed shape by invisible vacuum power was magical and fun to watch!

Today, I use vacuum forming to make toy prototypes in my own shop. I usually use a professionally made vacuum former, but in a pinch I've used this ultra-cheap, home-brew rig with great results.

Large, commercial machines have built-in vacuum pumps, adjustable plastic-holding frames, overhead radiant heaters, and pneumatic platens. The Guerrilla Vacuum Former is much simpler. It uses your oven to melt the plastic, and a household vacuum cleaner to supply the suction. All you have to build is a simple wooden frame and a hollow box. I'll show you how to do it, then use the device to create a tiki mask that also makes a great Jell-O mold.

Set up: p.109   Make it: p.110   Use it: p.115

Bob Knetzger is a designer/inventor/musician whose award-winning toys have been featured on *The Tonight Show*, *Nightline*, and *Good Morning America*.

Photography by Bob Knetzger

OPERATING IN A VACUUM

Vacuum forming is a two-stage process that's common in manufacturing.

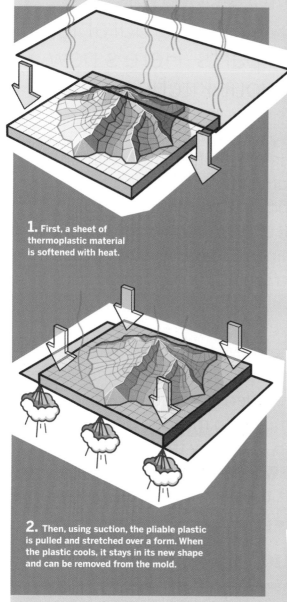

1. First, a sheet of thermoplastic material is softened with heat.

2. Then, using suction, the pliable plastic is pulled and stretched over a form. When the plastic cools, it stays in its new shape and can be removed from the mold.

The vacuum-forming process isn't something the public ordinarily sees, but it's amazing to watch. In the 1960s, Mattel's Vac-U-Form toy let kids actually run the process themselves. Compatible plastic sheets and mold sets were sold separately, of course.

The resulting parts are impermeable, uniform, and stackable, which makes the process ideal for mass-producing things like blister packs, coffee cup lids, cookie package trays, costume masks, store signs, protective panels, and raised-relief maps.

Illustrations by Damien Scogin

SET UP.

MATERIALS

[A] 2×4 lumber (1½"×3½") I used fir, which is cheap, easy to staple into, and, being wood, reasonably safe to handle when hot. You just need 4 short pieces, probably 2' or less each, so scraps are fine.

[B] Plywood or other material to make a shallow box. I had some scrap ½" plywood, but you can use particleboard, framing lumber, an old dresser drawer, a deep picture frame, whatever. In the spirit of guerrilla building, get creative!

[C] Pegboard 2'×2' should be enough

[D] Short piece of dowel or 2×2 lumber, or other small wood scrap

[E] Floor nozzle attachment for vacuum cleaner

[F] Drywall screws

[G] Duct tape

[H] Plastic sheets Polystyrene works well, but you can use nearly any thermoplastic material. 0.030" is a good thickness, or try thicker if you need a stronger part.*

TOOLS

[I] Shop-Vac or other vacuum cleaner

[J] Oven mitts

[K] Surface gauge (optional)

[L] Staple gun and staples

[M] Tape measure or measuring stick

[N] Saw

[O] Metal straightedge

[P] Drill bits (and drill)

[Q] Heat gun

[R] Plastic scribe (optional)

[NOT SHOWN]
Form material I used rigid urethane modeling foam, but you can use almost anything.

Oven with broiler

Woodcarving tools and a sharp knife

Sandpaper

Screwdriver

A few coins or a piece of window screen

*NOTE: Standard sheets are 4'×8', and a local plastic supply company may have odd sizes or scraps that will suit your project. IASCO-TESCO (iasco-tesco.com) also sells small sheets of styrene in various thicknesses and solid colors. I carved my form out of rigid urethane modeling foam, Last-A-Foam FR-7100 from General Plastics (www.generalplastics.com), but you can make a form out of almost anything (see sidebar on page 115).

MAKE IT.

BUILD YOUR GUERRILLA FORMER

START »» | **Time: 1 Hour Complexity: Easy**

1. MAKE THE FRAME

1a. Measure the interior of your oven, then subtract a few inches from the width and depth for clearance. This gives the size of the biggest frame you can make, which in turn determines the maximum size sheet of plastic you'll be able to mold. My oven is 21"×16", so I made an 18"×13" frame.

1b. Measure and cut the lengths of 2×4 you need to make your frame, allowing for the thickness of the sides. Make square cuts with your favorite saw: circular, jig, or handsaw. My frame called for two 18" pieces and two 10" pieces.

1c. Assemble the frame using 2 screws in each corner, for maximum strength. Drill pilot and clearance holes before fastening, and stagger the screws so as not to split the wood.

NOTE: The only critical feature of the frame is that it must lay flat with no gaps or high spots. This will ensure a good air seal when you vacuum form.

2. MAKE THE VACUUM BOX

2a. Calculate the dimensions of a shallow box with a footprint slightly larger than your frame; mine was 20"×15". The height should be enough to let you mount the vacuum nozzle attachment along one side. When you have the dimensions, cut 4 side pieces and a bottom out of plywood or other material, accounting for material thickness again.

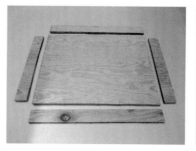

2b. Cut a top panel for the box out of pegboard, large enough to overlap all 4 sides. The pegboard's holes are used to suck air from under and around the form.

2c. Drill a large hole in the side of the box where you'll mount the vacuum floor tool. This is the main air vent.

2d. Use screws to assemble the 4 sides of the box, then add the back.

2e. Cut a piece of wood dowel or 2×2 to make a post that will support the center of the top panel. During vacuum forming, the top panel tends to bow in, and this support post prevents that, like one of those little plastic "pizza stacks."

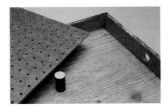
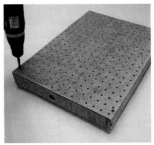

2f. Screw in the center post from the back of the box, then use screws to fasten the pegboard to the box sides.

2g. Mount the vacuum attachment over the side hole with small screws.

2h. Finally, seal all edges with duct tape. Don't forget the seams around the vacuum tool. Also tape a border around the top, leaving an open area in the center slightly smaller than your frame.

3. MAKE A FORM

3a. Plan your form. For my tiki head, I decided to use rigid urethane foam because it's easy to cut with woodworking tools, and my surfaces didn't need to be super smooth.

3b. Carve your form. I cut my tiki's profile using a band saw and made its basic shape with a small X-Acto saw. After sanding and filing the shape smooth, I carved in features and details with a Dremel tool, then used a set of model maker's "dental tools" for the smallest grooves and details.

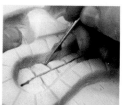

3c. If your form has concave parts, drill small holes completely through at the bottoms (at the surface's local minima).

NOTE: An extra-long 1/16" drill bit is very useful. The flutes are only on the end of the bit's long shaft, so you have to be careful. Drill only part-way, withdraw the drill to free the chips, then reinsert and drill some more. Repeat until you've drilled completely through.

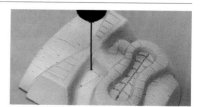

4. THIS PART SUCKS!

4a. Set up on the kitchen floor near your oven. Hook the vacuum cleaner up to the vacuum box nozzle, and test. The pegboard top should act like an air hockey table, but in reverse.

4b. Place your form onto the box. To help air flow under the mold, rest it on top of some coins as spacers. You can also put it on a piece of window screen.

> ⚠ **WARNING: Avoid using a table saw to cut plastic. If you must use one, install a special plastic-cutting blade.**

4c. Measure the plastic to just fit over your frame. Cut it down to size by scoring with a sharp knife, then bending it backward to snap along the score. Place the plastic over the frame and staple liberally. Don't skimp; the softened plastic will try to pull away, so you need to staple every inch or so.

4d. Place the frame in the oven, plastic side up, 4"–5" inches away from the broiler. Have your oven mitts ready, and set the broiler on high. Watch the plastic throughout its warming sequence, and be careful! First it will look wavy, then it will relax back flat, and finally the plastic will begin to sag in the center.

Test the plastic by pressing down near a corner with a pencil eraser or chopstick: if it's soft, it's ready to remove. If left too long under the heating element, the plastic will continue to melt, then smoke and burn!

4e. With your mitts on, remove the frame from the oven and flip it, plastic side down. Quickly and carefully place it over the form and press down. When the frame is pressed firmly flat against the top of the box, switch on the vacuum. Whoosh! The plastic is instantly sucked down over the form by vacuum power.

4f. If your part has some intricate detail or a hard-to-mold feature, use a heat gun to soften the plastic in that area.

Be careful, though — you could melt a hole right through the plastic.

4g. Continue pressing down on the frame for 20 more seconds with the vacuum on, until the plastic has cooled. Then turn it off and lift up the frame. Your form may fall out, but typically it will stay inside the formed plastic.

To remove the form, press on the back gently, or tap one edge of the frame firmly on a table or counter so that the form can fall away freely. You may even have to (carefully) pry the form out a little at a time with a screwdriver or knife.

5. TRIM AND FINISH

5a. Carefully remove the plastic from the frame. Look out for sharp staples!

5b. With a sharp knife or plastic scribe, score the plastic part to ready it for removal from the rest of the plastic. For a more precise trim, start by using a surface gauge to scribe a mark all the way around the part.

NOTE: If you need a lip on your part, trace around the shape, leaving a flat border all around.

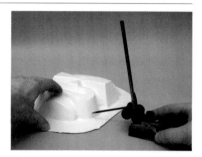

5c. Clean up the edges of the plastic with sandpaper, and trim as needed for your final application. That's it — you're done! But while you're at it, why not mold a few more?

FINISH ☒

NOW GO USE IT 》

FUN WITH MOLDS

Like any plastic mold, this mask has many applications, many of which happen to be party-related:
- Paint and add elastic to make the mask wearable.
- Use untrimmed, unpainted masks to make tiki-shaped Jell-O molds and punch bowl ice blocks for your next luau.
- Glue two masks back-to-back and install a bulb inside to make a glowing tiki lamp.
- Or enclose small speakers for an outdoor sound system.

What will you make with your Guerrilla Vacuum Former?

FORM MATERIALS AND PRACTICES

You can make your form out of nearly anything. This project uses urethane foam, but wood is another good choice: it's cheap; easy to drill, saw, and sand; and strong enough to form many parts.

For finely detailed forms, you can sculpt clay and then cast liquid plaster all around it. After the plaster hardens, remove the clay and drill air vents (see below). Parts made with this "female" mold will have details that match your clay original perfectly.

Found objects are a great source of shapes and details. You can add small plastic letters, caps and lids, or pieces of toys onto forms.

When designing a vacuum-form mold, there are a few guidelines to follow:

1. Avoid undercuts. Otherwise the plastic will wrap around underneath your form, trapping it.

2. Maintain some draft. For easier part release, avoid using vertical sides. Make your form with a few degrees of angle (draft) on all sides.

3. Drill vent holes. For concave details, drill a series

of tiny vent holes using a 1/16" drill bit. These allow the softened plastic to be sucked into the details.

4. Add height. If you'll trim your part away from the sheet, add a little extra height to the form, so that your part won't have any unformed edges.

SPIN THE BIRDIE

By Larry Cotton

BIRD SHOT

Birds make lousy subjects for digital photographs. They're fearful, fidgety, and, well, flighty. But you can improve your odds of getting awesome avian photos by moving your camera closer to the birds — and you farther away. And while you're at it, why not get them to pose for you?

Since converting from film to digital photography more than ten years ago (I bought an Epson PhotoPC with a half-megapixel resolution for $500 in 1996), I've sought ways to take advantage of the "tons o' shots to get a winner" phenomenon. Recently I discovered that bird photography definitely falls into that category.

But digitally shooting birds taxes a camera's resolution limit big-time: your 6 or 10MP digicam suddenly becomes another PhotoPC when you throw away valuable pixels by cropping. So, ya gotta get closer — but getting closer makes the birds more skittish and shy. It's a digital catch-22.

We *can* have our cake and eat it too, though, thanks to a long shutter-release cable and a gadget that typically controls model vehicles: an R/C radio.

Set up: p.119   Make it: p.120   Use it: p.125

Larry Cotton is a retired power-tool engineer, musician, part-time math teacher, coffee roaster, and bird harasser living in eastern North Carolina.

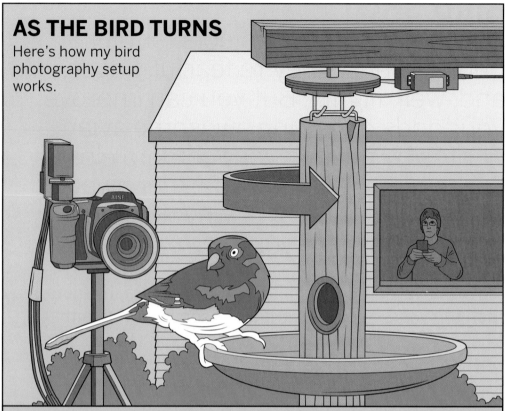

AS THE BIRD TURNS

Here's how my bird photography setup works.

To pose the birds, and to prolong camera battery life, I use an old R/C radio and its 2 servos: one actuates the camera's power switch, and the other slowly turns the bird feeder!

To control my camera's shutter release button, I use a long 3-conductor cable with 2 switches — the easiest, cheapest way to put your camera near the subject and shoot from a distance.

A shutter release button does 2 things: pressed halfway, it initiates metering and focusing; fully pressed, it fires the shutter. Some cameras have a jack for a shutter release cable, connected in parallel with the button; 2 switches at the other end are sequentially operable. The jack on my Pentax *ist DS camera accepts a standard $3/32$" (2.5mm) stereo plug, as do other DSLRs including the Canon EOS Digital Rebel, Samsung GX-1L, and Pentax K10D.

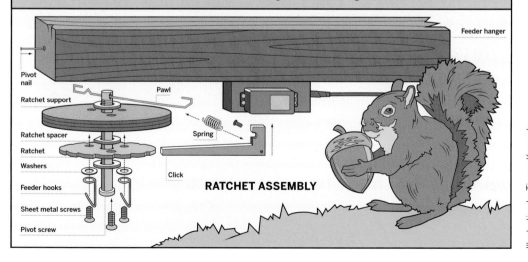

Feeder hanger

Pivot nail

Ratchet support

Pawl

Ratchet spacer

Ratchet

Spring

Washers

Click

Feeder hooks

Sheet metal screws

Pivot screw

RATCHET ASSEMBLY

Illustration by Timmy Kucynda

SET UP.

MATERIALS

[A] ½" plywood, 4"×4"

[B] Case for carrying it all. An old power-tool case works well.

[C] 3-conductor cable, at least 50' such as an old telephone cable; 6- or 8-conductor is even better.

[D] Sheet metal ¹⁄₆₄", several square inches from home supply store, or from a can or metal sign

[E] 24-gauge wire, several feet insulated or not

[F] 9V alkaline battery

[G] 6V lantern batteries (2) to power the R/C radio

[H] Small speaker such as All Electronics SK-214 or SK-285

[I] Laminated plastic 3"×3"×5" thick such as Formica

[J] Scrap wood

[K] Aluminum bar 1"×¾", about 14"

[L] PVC riser ½" internal diameter I used Home Depot 86106 (10 pack)

[M] Brass brazing rod ³⁄₃₂" diameter for feeder hooks. A coat hanger also will work.

[N] Large paper clip

[O] Pushbutton switches (2) I used All Electronics SMS-229, but any will do.

[P] Single pole single throw switches (2) Most any SPST switch works.

[Q] Electrical tape

[R] Acrylic sheet ⅛" thick from home improvement store or glazier, or other ⅛" material, for ratchet spacers

[S] Tower LXGRM7 R/C radio with 2 servos and receiver from Tower Hobbies (towerhobbies. com). Bigger servos have

more torque. Hardware varies according to configuration.

[T] BREADBOARD ASSEMBLY
• Photocell also called a photoresistor, RadioShack 276-1657
• IC breadboard I used RadioShack 276-175. Perf board is OK, too.
• 1K resistor RadioShack 271-1118
• 0.47µF capacitor All Electronics RMC-474
• 0.1µF capacitor RadioShack 272-1434
• 555 timer IC I used RadioShack 276-1718. CMOS also works.
• Small alligator clips (3–4)

[U] FASTENERS
• ¼" wire staple
• 4"–6" wire ties for wiring harness
• #16×1" wire nails (7)
• #16×1½" wire nails (4)
• ¼"-20×½" machine screw
• #10×¾" sheet metal screws (3)
• #10 flat washers (3)
• #8-32×3" machine screw

• #6×½" sheet metal screws (#5)
• #4×½" wood screws (2)
• #6-32×¼" machine screw and nut
• ¾" weak tension spring or small rubber band

[NOT SHOWN]
Stepladder

C-clamps (1 or 2)

Stereo plug 2.5mm (³⁄₃₂") from RadioShack or from yourcablestore.com. See page 118 for cameras that will take it.

Tripod

Camera DSLR preferred, but point-and-shoot OK

Solder

9V battery clip

Plastic O-ring or ¼M flat plastic faucet washer

Hanging bird feeder and bird seed

Music wire, .032" diameter

MAKE IT.

BRING YOUR
SUBJECT AROUND

START »

Time: A Weekend Complexity: Medium

1. MAKE THE SHUTTER-RELEASE SYSTEM

1a. Construct the cable. You can buy a ³/₃₂" (2.5mm) stereo plug from RadioShack. But if you're not into delicate soldering, spend a few bucks on an adapter cable (for example, yourcablestore.com part #HP 3M-2M 6) and cut off the ⅛" plug. Connect at least 50' of cable to the ³/₃₂" plug. An old telephone cable with at least 3 conductors works great.

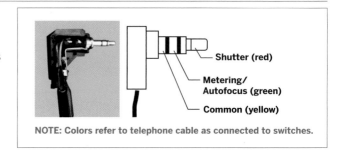

— **Shutter (red)**

— **Metering/ Autofocus (green)**

— **Common (yellow)**

NOTE: Colors refer to telephone cable as connected to switches.

1b. Mount switches on a switch block. The switch block is a noncritical block of wood; it just holds your switches. Any pushbutton (momentary-on) switches will work, but 2 micro switches with different actuating pressures can be combined into a sequential switch.

Nail (yes, nail) them to the switch block. With 2 nails, mount the one with the greater actuating pressure below the one with the lesser actuating pressure, which pivots around its own single mounting nail (don't pound it all the way down). Tap a small nail through the block from the back, to act as a stop for the pivoting switch. Wire as shown here.

1c. Your cable is now testable! Plug the stereo plug into your camera's shutter-release cable jack. Power up the camera. Press the first (or upper) switch, then press a little harder to actuate the second switch.

NOTE: Go to makezine.com/11/birdfeeder for tips on experimenting with your new shutter-release system.

2. BUILD THE CAMERA POWER SWITCH

2a. Make the switch coupling. My camera's main power switch rotates about 30 degrees and has a small lever-like protrusion. If yours is configured like that, Dremel-shape a small section of plastic plumbing riser (pipe) to match the camera switch on one end and the servo horn (the part that moves) on the other. Then screw the horn and your homemade coupling together with the #4×½" wood screws.

2b. Make the camera bracket. Mine is ugly and tortuously bent to hold the servo, receiver, and LED brackets. Make it from ⅛"×¾" aluminum extrusion. Be careful that the bracket doesn't block the optical viewfinder or the LCD; you'll need access to one of these for aiming and/or focusing the camera.

2c. Make the servo and receiver brackets. Made from thin sheet metal, these brackets attach the servo and receiver to the camera bracket. Use the assembly screws for mounting the servo and the #6-32 screw to mount the receiver.

2d. Mount the camera bracket loosely to the camera (for testing) by trapping its lower end between the camera and your tripod with a longer ¼"-20 machine screw, replacing the tripod's screw. Make sure the screw doesn't bottom out in the camera's tripod-mounting hole.

2e. Add the servo and receiver. Ensure the switch end of your coupling just snugly engages the camera's switch. Excessive pressure requires excessive torque (thus current) to turn the switch on and off.

2f. Test the servo coupling. With the ¼"-20 screw loosened and the coupling disengaged from the switch, connect and turn everything on. Press the appropriate joystick to observe how many degrees and in which direction the coupling moves. You can reposition the horn on the servo, tweak the trimmer on the transmitter, and/or limit the amount and direction of joystick travel by gluing a couple of pieces of thin plastic, such as Formica, near the joystick.

Photography by Sam Murphy

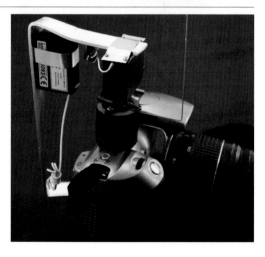

MAKE AN INDICATOR FLAG: Tighten the ¼"-20 screw and add a small but visible-from-a-distance "Camera On" flag to the servo's horn. Bend the music wire into an "L" shape and add colored tape or paper to the tip.

3. SPIN THE BIRDIE

3a. Make the feeder ratchet parts. The second R/C servo poses the birds by slowly ratcheting the feeder around. Make the ratchet parts by using the templates at makezine.com/11/birdfeeder (the rest are pre-made, from the Materials list on page 119).

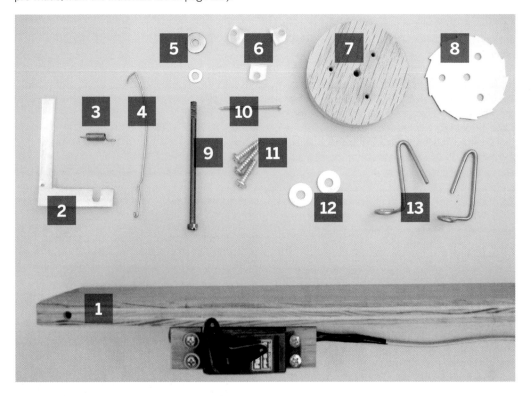

RATCHET ASSEMBLY

1. Feeder hanger Use a length of 3"×½" plywood to extend from your feeder support (pole, tree, etc.) to the feeder. Pre-drill appropriate holes in the other end to attach the hanger to your support.

2. Click Use thin ¹⁄₆₄" metal.

3. Spring

4. Pawl Use a paper clip.

5. Pivot screw washers

6. Ratchet spacers (3) made from ⅛" acrylic. Most any shape works as long as the hole is ¹⁄₁₆" from the curved edge. These must clear the pawl.

7. Ratchet support Use ½" plywood.

8. Ratchet Make it from ¹⁄₁₆" Formica using a band saw or a jigsaw mounted upside down; use a file to shape the teeth. The photo shows 14 teeth;

it's not critical as long as they're fairly evenly spaced and the throw on your servo is adequate to advance the ratchet 1 full tooth.

9. Ratchet pivot screw This pivot method using the 3" machine screw is almost foolproof. If you use a wood screw, make sure its threads are at least 1½" long.

10. Pivot screw nail

11. Sheet metal screws #10×¾"

12. #10 washers

13. Feeder hooks Since feeders vary, you'll customize these. Make them from ³⁄₃₂" brazing rod or a coat hanger. Bend an eye into one end of each hook for mounting to the ratchet. Use at least 2 hooks to smoothly transfer torque to the feeder.

NOTE: For parts templates, go to makezine.com/11/birdfeeder.

3b. Assemble the ratchet. Clamp the feeder hanger upside down in a vise. Using Figures 12 and 13 (online at makezine.com/11/ birdfeeder) as a guide, put together the ratchet assembly — except for the pawl and click.

3c. Mount the servo to the feeder hanger with small wood blocks. (Servo configurations may vary.)

3d. Add the pawl and click. Connect a short, weak tension spring or rubber band between them to keep them engaged with the ratchet.

3e. Connect and turn on the transmitter and receiver. Move the joystick, which actuates the feeder servo. Each pawl-pull should cause the click to engage an opposite-side tooth to prevent backward rotation. Each full cycle rotates the ratchet 24°–26°. You may have to adjust the action by slightly bending the pawl and/or click.

NOTE: To connect the ratchet assembly, extend the servo's wires so that they're longer than the distance from the feeder to the camera rig.

4. MAKE YOUR CAMERA FEEDBACK BEEPER (OPTIONAL)

Only one challenge remains: a way to signal "Picture taken!" without having the flash go off in Birdie's face.

On the back of my camera, there's a small LED that indicates memory-card access. A photocell (photoresistor) mounted face to face with this LED can change resistance in a simple 555 circuit (*see MAKE, Volume 10, page 62, "The Biggest Little Chip"*), creating or changing a tone in a speaker.

If your camera has this LED, make and mount another small bracket of thin sheet metal to bridge between the camera bracket and the LED. Drill a hole to fit the photocell, while holding it in good contact with the LED. Wrap black electrical tape around the assembly (don't cover the face of the photo-cell!) and put a small O-ring around the photocell to shield it from extraneous light.

Build the beeper per the schematic shown online, using proto-type or perf board. Mount it near the R/C transmitter, then run 2 wires from the photocell, alongside the remote shutter-release cable, to the beeper.

NOTE: For the schematic on the feedback beeper, go to makezine.com/11/birdfeeder.

5. MOUNT REMOTE CONTROLS AND WIRE IT UP

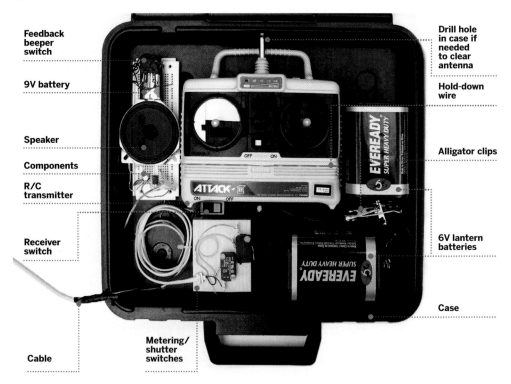

Feedback beeper switch

9V battery

Speaker

Components

R/C transmitter

Receiver switch

Cable

Metering/ shutter switches

Drill hole in case if needed to clear antenna

Hold-down wire

Alligator clips

6V lantern batteries

Case

Since you're running wires anyway, run another pair to power the receiver.

I mounted all the remote stuff in an old plastic power-tool case using hot-melt glue and wire pulled through holes in the case to hold everything in place. You can also glue foam blocks to the inside of the top.

Mount the meter/shutter switch block near the camera-on joystick, for one-hand operation. You'll need the other hand to move the feeder-turning joystick.

Three more switches turn everything on: 1 in the transmitter, 1 for the receiver, and 1 for the feedback beeper. The 2 you add can be any SPST switches.

One 9V and two 6V (lantern-size) batteries power everything. To replace the (usually) 8 AA batteries in your transmitter, solder wires to the first (+) and last (-) battery terminals, run them out of the housing, and connect them to the lantern batteries, in series, with alligator clips. Wire the whole rig according to the diagram online at makezine.com/11/birdfeeder.

FINISH

NOW GO USE IT »

USE IT.

GET A BIRD'S-EYE VIEW

NOTE: If you're using the camera feedback beeper, you'll hear a beep (or a pitch change) from the speaker when you take a picture.

Set the camera up and move away from it, as in the shutter-cable testing procedure.

Turn on the transmitter, receiver, and feedback beeper (if using). Move and hold the joystick to turn the camera on. (Watch the flag on the camera for confirmation.) While holding the joystick on, actuate the metering and shutter switches.

Now practice rotating the feeder by repeatedly moving its joystick s-l-o-w-l-y.

Take another shot or two, then retrieve your memory card and download the picture(s) to your computer to check focus, exposure, composition, etc. (Unless you're extremely lucky, you'll probably just have beautiful shots of your bird feeder.)

Finally, reinsert a formatted memory card, get comfortably away from the feeder, and hold the plastic case in your lap. Don't turn on the camera until a bird lands on the feeder.

Using both joysticks and the meter/shutter switches, take lots of pictures while you slowly rotate the feeder to the best camera angle. You may also turn the feeder between bird visits; they may prefer certain feeding positions to others.

Do the birds like to be rotated? Eventually, most of them adapt. Some are more skittish than others; some actually seem to enjoy it. The female red-wing blackbirds stop eating and look up at the sky. Blue jays bolt. And cardinals just smile.

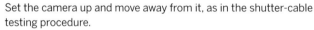

Easy Motor
By Cy Tymony

Make a spinning motor with a minimum of parts.

You will need: Two strong, metallic disc magnets approximately ¾" in diameter (larger in diameter than a AA battery), 6" length of stiff copper wire, AA battery, needlenose pliers

1. Make it.

First, if your copper wire is insulated, strip the insulation off. Bend the top of the wire into a hook shape. Press the tip into a sharp point with the pliers. Wrap the wire in a spiral form around the body of the battery, as shown in Figure 1. The wire should be just loose enough so it will not contact the battery case.

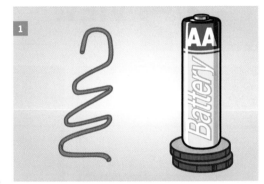

Next, place 2 metallic disc magnets on the battery's negative terminal. Adjust the shape of the wire so the top rests on the center positive terminal and the bottom just touches the side of the magnet. The length, when twisted, from the tip of the top hook to the bottom of the wire should be 2".

2. Try it.

Once the wire makes contact with the top battery terminal, it will spin. If it does not, carefully bend and adjust the wire so that it is free to move and make contact properly. To make the wire spin in the other direction, turn the magnets over.

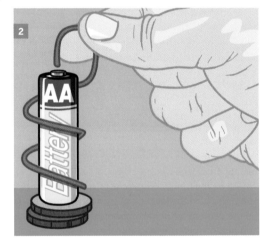

3. Understand it.

When the wire touches the battery's positive top terminal and the side of the magnet, electricity flows through it. The wire becomes magnetic and is repelled by the 2 disc magnets. When the wire moves away from the magnet, it disconnects from the battery and loses its magnetic field. It falls back to its original position and contacts the side of the magnet. Then, it connects to the battery through the metallic magnet and becomes magnetically charged again, and the cycle repeats.

Since the wire coil is suspended by its sharp tip on top of the battery's positive terminal, when it's repelled from and returns to the magnet, it rotates slightly. The cycle occurs so rapidly that it produces a circular motor motion.

Cy Tymony is the author of the *Sneaky Uses for Everyday Things* book series. sneakyuses.com

Illustrations by Dustin Hostetler

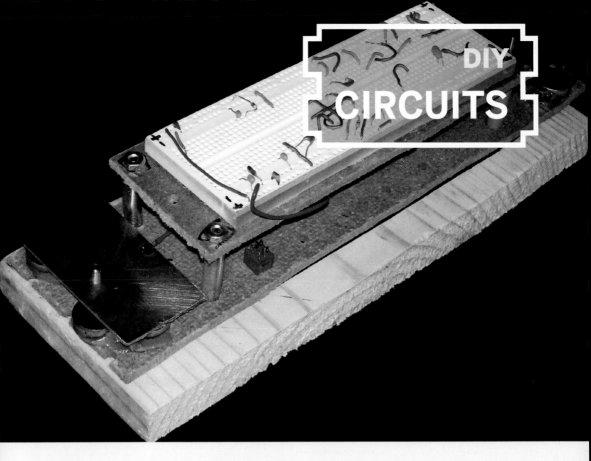

BREADBOARD RACK

A simple way to interconnect protoboards.
By Tom Zimmerman

Solderless protoboards are a convenient way to build electronic circuits, but interconnecting several boards gets messy. Adding input and output (I/O) components such as switches, potentiometers, jacks, and LEDs requires drilling, wiring, and stuffing it all into a box, where it's hard to debug and modify.

A nicer approach, which I borrowed from early analog synthesizers, is to build handy modules that carry one protoboard each and plug into a rack. Each module carries its protoboard underneath, and the circuit's I/O components are on a visible and easily reachable front panel, just like any other device that hides its complex circuitry inside and exposes only the controls and connectors.

The front panels are made of pegboard, providing predrilled guide holes for accurate alignment of module components. The rack holds 14 modules,

7 on each side. Wider modules can be built to accommodate larger protoboards. Angle iron rails both hold the modules and supply power. The top rail supplies modules with unregulated +12 volts, and the bottom rail supplies electrical ground. Magnets behind copper contacts on each module grab the rail, hold the module in place, and provide a large surface area to deliver plenty of current. Each module must have a power diode, for inevitably a module will be accidentally installed upside down. Additionally, a voltage regulator on each module assures a clean and stable power supply.

Before you start, remember these important safety tips: wear eye shields when working with power and hand tools, gloves when cutting metal and applying hot glue, and a dust mask when sanding and cutting.

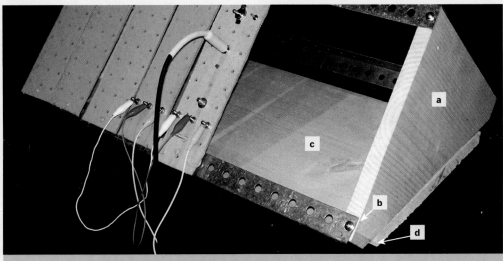

Fig. A: Rack construction.

MATERIALS

1¼"×24" zinc-plated, punched angle iron (3)
Make sure it contains iron: test with a magnet.
1×12 pine board, 52" length **(It's actually 11⅜".)**
4'×2' sheet of pegboard **for module faceplates**
¾" magnets (112 total), 8 per module **the stronger, the better**
Copper-clad fiberglass (126 in²) **for the rail contacts**
1½" 8-32 screws (100) **That's 6 per module (2 to grab rails and 4 for protoboard standoff) plus extras for interconnection posts.**
8-32 nuts (100) **for all the 8-32 screws**
1⅝" #8 sharp lath screws (10) **for attaching rail to wood, and wood to wood**
³⁄₁₆" inner-diameter vinyl tubing (6') **for standoff material**
2¼"×6½" solderless protoboards (14)
RadioShack #276-174
1N4001 diodes (14) **RadioShack #276-1101, prevents reverse voltage from module installed upside down.**
Masking tape
Steel wool **or fine emery cloth**
Black and red insulated wire
12V power supply

TOOLS

Jigsaw and hacksaw
Hot glue gun
File
Soldering iron and solder
Drill with drill bits:
 ³⁄₃₂" twist bit **for the pilot hole for lath screw**
 ⅛" twist bit **for slip hole for lath screw**
 ³⁄₁₆" twist bit **for pass-through hole for 8-32 screw**
 ½" boring bit **to countersink lath screws**

Triangular Side Dimensions

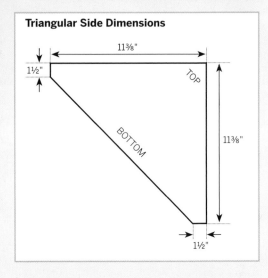

11⅜"

1½"

TOP

BOTTOM

11⅜"

1½"

1. Build the rack and power the rails.

1a. Cut the angle iron with the hacksaw to make 3 pieces, each 2' long. File the corners for safety. Then cut 2 triangular sides out of pine board following the diagram above.

1b. Assemble the prism-shaped rack. Mark drill holes on the wooden sides (Figure A, label a) so that the rails end ⅛" in from the edge, where they won't stick out and scratch people (label b). Drill one ³⁄₃₂" pilot hole for each rail end (2 screws would collide in the wood). Screw the 3 rails to the triangles. Place a stripped wire under one end of each rail before screwing it down. Use black wire for the bottom rails

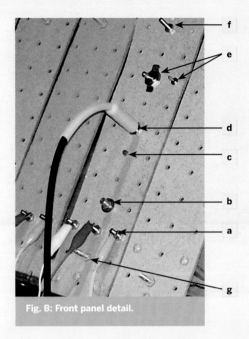

Fig. B: Front panel detail.

pair of angle iron holes. Trace the angle iron holes onto the faceplate and drill a ³⁄₁₆" hole for each. This is the only time you will drill the pegboard where it isn't pre-drilled. Install a 8-32 screw in each hole and secure with a nut. These screws will keep the module aligned with the rails (Figure C, label a).

2b. Decide what I/O components you want for your module (Figure B), such as a potentiometer (b), LED (c), ¼" phono jack (d), and large and micro switches (e). A row of 8-32 screws will act as posts for interconnecting modules using alligator clips (a).

2c. Mount the components into the holes of your pegboard faceplate, enlarging holes where necessary. A Unibit (a single drill bit with stepped diameters) works well for this. Also install two 8-32 screws centered one hole in from each end, and pointing outward with the nut on the faceplate front (Figure B, labels f and g). These "module grabber" screws provide a convenient way to pull modules out and hold them in a stand during circuit assembly.

2d. Cut a protoboard backing out of pegboard, measuring 2⅝"×7⅝" and with 8×3 complete holes. Then hot-glue the solderless protoboard to the backing pegboard, centering it so the top and bottom rows of the pegboard holes are accessible. Once the protoboard is glued, you cannot remove it without damaging it, so position it carefully.

2e. Install four 8-32 screws on the faceplate in a rectangle: 2 holes apart, 2 holes from the top, and 3 holes from the bottom. Cut four 1⅛" pieces of vinyl tube and slip them over the screws to create standoffs (Figure C, label b), then fit the protoboard backing onto the screws and fasten down with nuts.

and red for the top. Connect the black wires to the ground of a 12V power supply (sometimes labeled "-"), and the red wire to the +12V.

1c. Cut a 24" piece of the 1×12 wood and screw it to the bottom of the rack, piloting the holes with a ³⁄₃₂" bit and using the ½" bore bit to countersink (Figure A, label c). The gap between the rack bottom and triangle sides will provide ventilation (label d).

2. Make a circuit module.

2a. Cut a faceplate out of pegboard, measuring 2⅞"×11⅝", so that there are 11×3 complete holes. Lay this faceplate across the top and bottom rails, so that it overlaps the rails equally, and the center column of holes on the faceplate lines up with a

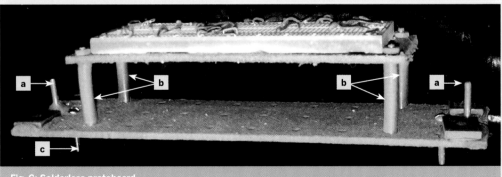

Fig. C: Solderless protoboard.

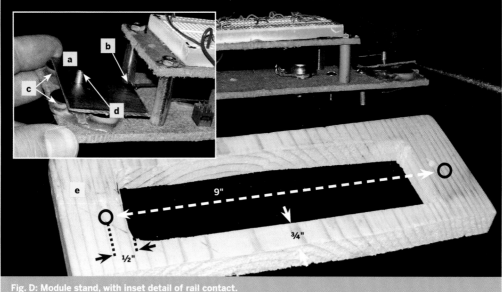

Fig. D: Module stand, with inset detail of rail contact.

3. Attach the rail contacts.

3a. Cut 2 pieces of copper-clad board 2⅝"×1⅝" for the rail contacts (Figure D, label a). Drill a ³⁄₁₆" hole centered along the width, ⅜" in from the long edge. Polish with steel wool or fine emery cloth until the copper is shiny. Solder a 12" wire to the copper edge opposite the hole, to deliver power from the rail (Figure D, label b). Use red wire for the positive contact at the top of the module and black wire for the ground contact at the bottom.

Connect the other end of the red wire to the anode (non-striped) side of a 1N4001 diode. Push the other, cathode (striped) end of the diode into the bus strip on the protoboard that you'll use as your circuitry's positive power supply. Push the other end of the black wire into the bus strip that you'll use as ground. The diode protects your protoboard's circuitry if you accidentally place the module upside down on the rails.

3b. On the back of the module faceplate, place 4 magnets at each end, within the rail contact area (Figure D, label c). Stick masking tape down onto the magnets, then use the tape to pull them off the pegboard. Apply hot glue to the pegboard rail contact area. Holding only the tape, push the magnets back into place on the glue. Make sure they are flush and don't rest on the nut of the alignment screw. Once the glue hardens, remove the tape. The tape prevents the glue from getting everywhere.

3c. Make sure each copper contact fits over the rail alignment screw (Figure D, label d). Hot-glue each contact to the magnets. Your module is ready!

4. Make a module stand.

4a. When you are working on a module, the I/O components and grabber screws will prevent the module from laying flat on a table, hence the need for the module stand (Figure D, label e).

4b. Cut a piece of 4"×11⅜" pine, and use pegboard as a template to mark 2 holes centered 9" (9 holes) apart. Drill ³⁄₁₆" holes completely through the wood.

4c. Draw a rectangle on the block ¾" from the sides and ½" in from the holes. Drill a ½" hole 2" from the corner. Clamp the block in a vise, insert the jigsaw blade in the hole, and cut out the rectangle. Curve the corners slowly, then square them up later so you don't snap your jigsaw blade. Sand all edges to remove potential splinters. Your stand is done!

To use the stand, place the module facedown with the grabber screws fitting into the 2 holes in the stand. Clip a battery to the copper rail contacts to power the protoboard for testing and debugging.

Tom Zimmerman is an inventor, educator, and researcher at the IBM Almaden Research Center who loves gadgets, LEDs, synthesizers, and hooking people up to computers.

BIG KID NIGHT LIGHT

 Dollar-store hack makes a color-changing mood light. By Dan Weiss

Kids want night lights, they just don't want them to look like night lights because that makes them "babies." Here's a shining star made from things you can pick up at a dollar store. Call it a "mood light," and no one has to admit that they use a night light.

1. Open the touch lamp by removing the screws in the back with a Phillips screwdriver. Be careful that you don't lose any of the springs inside.

2. Use a utility knife to cut the wires leading to the light bulb (Figure A). Cut as close as you can, so you have the longest wires to work with. Remove the bulb from the lamp. Strip the ends of the wires.

3. You'll use the switch built into the touch lamp and the power from the negative side of its battery compartment. In this lamp the negative wire is blue (and not very long). You need to cut this wire in half (making it even shorter) and strip the ends. Be careful when pulling on this wire. If the solder connecting it to the battery case is weak, the wire will break off.

4. Now let's turn to the LED ornament light. Flip the light over and use the flathead screwdriver to pry off the cover. Remove the holographic foil and save it for later. Using the flathead screwdriver, gently pry the LED module off the 2 posts (Figure B, circled in white). There's no glue, so it should be easy. Cut all 4 wires, including the ones going to the switch, with the utility knife as far from the module as possible so you have the most wire to work with (Figure B). Strip the ends of all the wires.

A

B

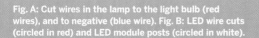

C

D

Fig. A: Cut wires in the lamp to the light bulb (red wires), and to negative (blue wire). Fig. B: LED wire cuts (circled in red) and LED module posts (circled in white).

Fig. C: Wire the LED module into the touch lamp, as shown. Fig. D: Optional light diffuser — a plastic bottle cap glued over the LEDs.

MATERIALS

Dollar store touch lamp in your choice of shape
Dollar store LED ornament light
   aka color-changing LED light base
AA batteries (4)
Screwdrivers small Phillips head and flathead
Utility knife

OPTIONAL

White plastic bottle cap as from a bottle of water
Glue pretty much any type
Soldering iron and solder
Electrical tape

5. Now we Frankenstein these 2 items together (Figure C). Place the LED module in the center of the lamp where the bulb sat. Connect the module's yellow wire to the red wire from the lamp's battery case. Connect the module's blue wire to the blue wire from the battery case. These are the power wires.

Connect the 2 black switch wires from the module to the blue and red switch wires on the lamp. You can connect the black wires either way.

Push the wires toward the LED module, not touching each other, and cover them with the holographic foil from the LED ornament light. This adhesive plastic foil will hold the wires in place.

You can cover the wire connections with electrical tape and even solder them, but it's not required.

6. This step is optional, but I found that the LEDs shined too sharply through the lamp, so I used a white plastic bottle cap as a diffuser. Glue the cap to the holographic foil over the LEDs (Figure D).

7. Close the lamp up and install the batteries. When assembling the lamp, be careful that the posts are in the springs, otherwise the lamp will not "bounce back" like it should.

The original LED lamp took 3 AA batteries and the new lamp uses 4, so we're giving the LEDs more power than intended, but not that much more.

You're done! When I finished this lamp, my daughter immediately claimed it as her own and she's been using it ever since.

▶ To see the lamp in action, go to makezine.com/11/diycircuits_nightlight.

Computer geek Dan Weiss resides in steamy St. Louis, Mo., and finds a particular joy in mashing up dollar-store items to create useful and whimsical artifacts.

Photography by Dan Weiss

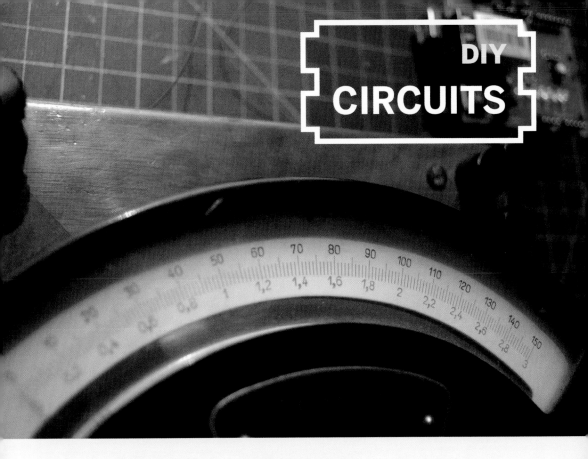

NET DATA METER

Antique voltmeter displays current air quality from the web. By Tom Igoe

One thing that disappoints me about computers is how little character they possess. Antique instruments of information display, like Victorian pendulum clocks, barometers, and compasses, and Babbage's calculating engines, have a presence that modern computers lack.

I dig the look of the iPod as much the next guy, but even the best manufacturing design today doesn't match that old brass-and-hardwood handcrafted love. Desktop widgets replace the need for clocks, barometers and stock tickers, and multipurpose display hardware like the Ambient Orb also perform these functions. But because these things have little presence and are so easily reconfigured, it's easy to forget what information they're displaying. Does the meter's sudden plunge mean my Google stock tanked, or that it's going to rain tomorrow?

Many geeks, myself included, resist this trend by fanatically collecting old instruments for aesthetic reasons. I've been playing around with using antique instruments to display data from new sources. I have some beautiful, wood-encased voltmeters and ammeters from the early 20th century, which I rescued from my university's physics department trash. They work quite well but they're not as portable as my current multimeter, so I wasn't using them, which is a tragedy for such proud, functional instruments. So I decided I'd take a cue from Ambient Devices (ambientdevices.com) and make one of them into an air quality monitor.

First, I came up with the basic system. The meter I used is an analog voltmeter that ranges from 0–3V DC. That's a good range to control from a microcontroller, so I decided to use my

Photography Tom Igoe

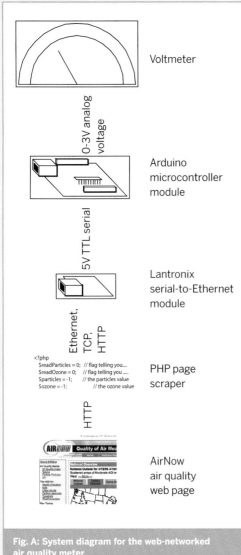

Voltmeter

0-3V analog voltage

Arduino microcontroller module

5V TTL serial

Lantronix serial-to-Ethernet module

Ethernet, TCP, HTTP

```
<?php
$readParticles = 0;   // flag telling you....
$readOzone = 0;       // flag telling you .....
$particles = -1;      // the particles value
$ozone = -1;          // the ozone value
```

PHP page scraper

HTTP

AirNow air quality web page

Fig. A: System diagram for the web-networked air quality meter.

Figure B: Pulse Width Modulation

Voltage / Time

Apparent voltage

Duty Cycle = 50 %
(on half the time, off half the time)

Voltage / Time

Apparent voltage

Duty Cycle = 20 %
(on 20% of each cycle, off 80% ofeach cycle)

Fig. C: AirNow's page is nicely laid out for scraping.

current favorite microcontroller module, the Arduino (arduino.cc). Setting up an air quality monitoring station seemed like more work than I wanted to do, but fortunately the data for local areas all over the United States is available online.

AirNow (airnow.gov) is an interagency website that reports the Air Quality Index (AQI) from local monitoring stations nationwide. I would connect the microcontroller to the internet using a Lantronix serial-to-Ethernet module (lantronix.com), then extract the data from AirNow's page for my city, New York City.

Pulse Width Modulation

The system layout made sense (Figure A), but the microcontroller needed to send a varying voltage to the meter, and microcontrollers can't output analog voltages. They can, however, generate a series of very rapid on-and-off pulses that can be filtered to give an average voltage. The higher the ratio of on-time to off-time in each pulse, the higher the average voltage (Figure B). This technique is called *pulse width modulation* (PWM).

For a PWM signal to appear as an analog signal, the device receiving the pulses has to react more

Fig. D: The antique voltmeter, attached to the Aruino microcontroller.

slowly than the pulse rate. For example, you can pulse width modulate a dimming effect for an LED because the human eye can't detect on-off transitions faster than about 30Hz.

Analog voltmeters are slow to react to changing voltages, so PWM will also work well for our antique display. To drive the meter, I would connect its positive terminal to an output pin of the microcontroller and its negative pin to ground. Then I could control its reading by pulse width modulating the micro's output pin.

1:1 Scaling

The meter's display reads from 0 to 150, and the AQI runs from 0 up to 500. But the EPA regards air that measures 150 as "unhealthy for all individuals," so I decided to set my meter up to show the raw AQI, without any scaling. If I see the needle pegged at the high end, I'll know I just shouldn't breathe.

Parsing the AirNow Page

The next step was to get the data from AirNow's website into some form that the microcontroller could read. The microcontroller can easily read in short strings and convert the ASCII into binary, but it would be tough to parse through all of the text on

a web page and find the right string. So I decided to write a program on my own server that would parse the AirNow page, extract just the current AQI reading for New York, and save it someplace where the microcontroller could read it. The microcontroller could then request a TCP connection via the serial-to-Ethernet converter and read in the data.

AirNow's page is formatted well for extracting the data (Figure C). The AQI number is shown clearly in text, and if you remove all of the HTML tags from the page source, it always appears on a line by itself following the line "AQI observed at hh:mm AM/PM:". I wrote a short PHP script to read the page, strip out the HTML, and find those two lines. When it did, it returned the AQI value by itself like so:

```
< AQI: 43>
```

On my server, a cron job runs the PHP script periodically and writes the value returned into a file that's accessible via Hypertext Transport Protocol (HTTP).

Serial-to-Ethernet Interface

The next step was to connect the microcontroller to the serial-to-Ethernet converter and then to the net. I used the Xport device from Lantronix, which makes it easy. Like other Lantronix devices, the Xport has a TCP/IP stack and simple web and telnet interfaces built in on the Ethernet side. On the serial side, it uses the same TTL serial protocol as most microcontrollers, including the Arduino, so hooking them up means simply connecting the micro's transmit line to the converter's receive line and vice versa (Figure E). I used an Xport for this project because I had a custom-printed circuit board designed for it, but if you're new to these devices, you might want to start with the Lantronix Micro. The Micro has a simpler connector and can be wired to a solderless breadboard with an IDE connector and a ribbon cable.

Before you can connect the Lantronix device to the net, you have to configure it. Lantronix has a downloadable configuration utility for Windows, DeviceInstaller. For non-Windows users I have a couple of programs that will do the job (one for Java, one for Processing) available online at makezine.com/11/diycircuits_meter.

On a network with DHCP enabled, the Lantronix devices will obtain an address automatically. Once you know the device's address, you can telnet into it to configure its serial port and network settings. Here are the settings I used for this project:

```
*** Basic parameters
IP addr 192.168.0.23, gateway 192.168.0.1, netmask
255.255.255.000 (8 bits)
*** Channel 1
Baudrate 9600, I/F Mode 4C, Flow 00
Port 10001
Remote IP Adr: --- none ---, Port 00000
Connect Mode : D4
Disconn Mode : 00
Flush   Mode : 00
```

Communications and Code

The microcontroller then connects to a web server by sending the Lantronix device a connect string that specifies the numerical address of the server and the port number:

```
C204.15.193.131/80
```

Once a connection is made, the Lantronix device returns a "C" to confirm. After that, any data sent in either direction passes right through between microcontroller and server, as through a serial port connection.

The full Arduino code for my microcontroller is online at makezine.com/11/diycircuits_meter. It connects to the net with a method like this:

```
void xportConnect() {
  //   send out the server address and
  //   wait for a "C" byte to come back.
  //   fill in your server's numerical address below:
  Serial.print("C204.193.131/80");
  status = connecting;
}
```

Then it waits for the Lantronix device to return with a "C":

```
if (status == connecting) {
  // read the serial port:
  if (Serial.available()) {
    inByte = Serial.read();
    if (inByte == 67) {  // 'C' in ascii
    status = connected;
    }
  }
}
```

Once it's connected, it sends an HTTP request like this:

```
void httpRequest() {
  //  Make HTTP GET request. Fill in the path to your
version
  //  of the CGI script:
  Serial.print("GET /~myaccount/scraper.php HTTP/1.1\
n");
  //  Fill in your server's name:
  Serial.print("HOST: www.myserver.com\n\n");
  status = requesting;
}
```

And the server replies:

```
HTTP/1.1 200 OK
Date: Fri, 14 Apr 2006 21:31:37 GMT
Server: Apache/2.0.52 (Red Hat)
Content-Length: 10
Connection: close
Content-Type: text/html; charset=UTF-8
< AQI: 65>
```

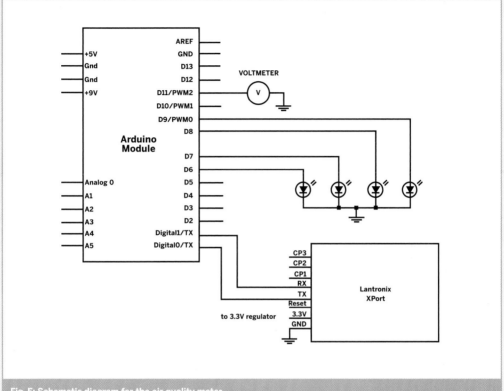

Fig. E: Schematic diagram for the air quality meter.

When you call this PHP script from a browser, you don't see the header stuff at the top because the browser strips it out for you. In the Arduino program, I stripped the header out by ignoring all the bytes before the < sign. Then I took only the numeric characters from the remaining string, converted them to a binary value, and I had my Air Quality Index value.

The final step was to pulse width modulate the meter. This is simple in Arduino, using the analog Write command:

```
void setMeter(int desiredValue) {
  int airQualityValue = 0;
  // if the value won't peg the meter, convert it
  // to the meter scale and send it out:
  if (desiredValue <= meterScale) {
    airQualityValue = desiredValue * meterMax /meterScale;
    analogWrite(meterPin, airQualityValue);
  }
}
```

Das Blinkenlights

As a finishing touch, I added LEDs to 4 of the Arduino's digital outputs so I could monitor the progress of the connection. LEDs hanging off of the Arduino's digital I/O pins 6–9 indicate the status Disconnected, Connected, Connecting, and Requesting, respectively.

That's all there is to it! The only remaining step is to build a false base for the meter that houses the electronics. Once that's done, you've got an attractive way to track local air quality, and you've put a well-made instrument back into useful service.

Tom Igoe teaches physical computing and sustainable technology development at the Interactive Telecommunications Program (ITP) at NYU. He hopes one day to work with monkeys.

FETCH THE WEATHER WITH THE MAKE CONTROLLER

This easy starter project displays your local forecast. By Brian Jepson

In the expanding universe of microcontroller boards, the MAKE Controller Kit fills the space between the easy, cheap Arduino and the more complex, powerful Gumstix devices. The low-power Arduino runs 8-bit machine code and can only do one thing at a time. The Linux-based Gumstix can do what a Linux-powered PC can do.

The MAKE Controller has the advantages of both, thanks to its FreeRTOS operating system; it can run multiple tasks simultaneously but also lets you allocate processor time more explicitly than you could under Linux. Because you need to know C, it's a little more complicated to program than Arduino, but still easier to use than the Linux-based Gumstix.

Making the Physical Connections

Once I got my hands on a MAKE Controller, I saw that it can talk to things in my house with its serial port, and talk to things on the internet with its Ethernet port. My first project was simple: download the Weather Underground's RSS feed for my hometown, yank out the forecast, and display it on a little LCD.

You need to solder your own headers or jack to the MAKE Controller's serial port contacts, a group of 6 holes, marked 3.3V, 0V, TX, TRX, RTS, and CTS. To be able to talk to my Serial LCD, I only had to make 3 connections: power (3.3V), ground (0V), and serial transmit (TX).

Photograph by Corbis

5 volts of USB power from a computer flows into the MAKE Controller (1), plus an extra 5 volts from a MintyBoost or similar power source (2), while VExt1 is set to the center position (3)

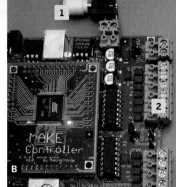

9 volts flows into the MAKE Controller from an external power supply (1), while VExt1 is set to the 5V position (2)

Fig. A: You can power the MAKE Controller and LCD using 2 sources ...
Fig. B: ... or just 1 source.
Fig. C: The mchelper program lets you upload a new program to the MAKE Controller, send messages to it, and monitor what it's up to. Fig. D: MAKE Controller reading the weather. Remember, if you get angry about the forecast, don't take it out on your MAKE Controller!

KNOW YOUR SERIAL CONNECTIONS

Don't confuse RS-232 and TTL serial interfaces. RS-232 is an old standard for connecting a PC and a modem, mouse, or Palm handheld. TTL (transistor-transistor logic) is a different beast that defines how small devices like the MAKE Controller and Spark Fun serial LCD talk to each other. With TTL, the voltages are lower and the signals are different. If you connect an RS-232 device to a TTL device, it won't work, and you may fry the TTL device.

Avoid confusion by reading spec sheets or using a voltmeter to check signal voltages. From 3V to 5V is usually TTL; -3V to 15V is RS-232. If you need to connect RS-232 and TTL devices, use a level shifter such as Spark Fun part number PRT-00133.

For more info, or to buy a MAKE Controller Kit, a MintyBoost kit, or various other kits, visit the Make:it (kits) section at store.makezine.com.

Unfortunately, the 3.3V wasn't enough to power my LCD (Spark Fun part number LCD-00461), so I needed an external power source. The MAKE Controller lets you handle this in a couple of ways.

One is to feed 5V, like from a Minty Boost (ladyada. net/make/mintyboost), into the board via its VExt1/Gnd pins, and set the voltage select jumper to the VExt position, to pass the voltage through to the LCD. Another is to connect 9V battery power to Gnd/V+ on the main power connector and set the jumper to the 5V position, to step the voltage down.

With the 5V input to VExt1, you also need to connect a USB cable in order to power the board itself. A 9V source gives enough power for both controller board and peripheral display, so you don't also need USB.

No matter how you decide to put it together, you'll power the LCD with the MAKE board's VOut1 and Gnd, and connect the board's TX transmit to the LCD's RX receive. The serial LCD must share a ground with the board, which is why you can't just power it on its own.

Programming the MAKE Controller

Programming the MAKE Controller is easiest if you've already worked in C. For a jump-start on the C language, check out *Practical C Programming, Third Edition* by Steve Oualline. After that, I recommend Peter van der Linden's *Expert C Programming: Deep C Secrets*.

To start programming the MAKE Controller, download and install the source code and tools from the MakingThings SourceForge project page: sourceforge.net/projects/makingthings.

You'll need 3 things:
- The source code for the firmware
- The GNU ARM tools
- mchelper

Installing the tools can be a little complicated. For help, check out the tutorial "Programming the MAKE Controller" at makingthings.com/resources/tutorials.

Once you've installed everything, try building the "heavy" project that comes with the firmware. To do this, open up a terminal (on Mac OS X or Linux) or Command Prompt (on Windows), change directory to the *heavy* subdirectory of the firmware source, and type the command `make`. The code should compile and create the file *heavy.bin* in the *output* subdirectory. Upload this file to your MAKE Controller using the mchelper program as follows:

- Open up mchelper, click Browse, and select the *heavy.bin* that you just created.
- With the MAKE Controller plugged into your computer's USB port, short out the ERASE pins on the MAKE Controller application board.
- Unplug the MAKE Controller and plug it back into your computer.
- Quickly (don't wait too long or the existing program may start running) click the Upload button in the mchelper application.
- Following mchelper's "Upload complete — reset the power" prompt, unplug and plug in the MAKE Controller again.

For more help, check out the resources at the makingthings.com website, especially the discussion forum. If you're into Internet Relay Chat (IRC), be sure to check out #makingthings on irc.freenode.net. You'll find many of the folks from MakingThings there, and I lurk there from time to time myself.

If you've made it this far, you're ready to start hacking your own code on the MAKE Controller.

Making the Weather Reader

I won't make you look at (or type in) the entire RSS Weather Reader program, but you can download it at makingthings.com/projects/rss-weather-reader.

The Weather Reader code is based on the *heavy* program that you just compiled. Make a copy of the *heavy* source code directory and rename it *Weather Reader*. Once you've done this, swap in my version of the file *make.c* and modify it to suit your needs.

In the definitions and declarations at the top, change the `MY_IP` and `GATEWAY` definitions to match your network's IP and gateway (router) addresses. (Yes, those are commas in between the byte numbers instead of periods — that's because these values are passed as 4 separate parameters to one of the functions.)

The weather fetching and display action happens in the `SerialTask()` function. First, it turns on the serial port at 9600 baud so it can talk to the LCD. Next, it connects to the Weather Underground at `64.243.174.104`, which is the numeric address for rss.wunderground.com. It requests the weather forecast using a `GET` request of this form: `/auto/rss_full/RI/Kingston.xml?units=both`. Change this call in your code to match your location. You can type the string into a browser to see the raw RSS feed and make sure you have the correct resource location.

`SerialTask` chews on the RSS feed, stores the weather forecast into a message buffer, then displays it on the LCD screen. After it's displayed it 10 times, it downloads and displays the weather forecast again. Lather. Rinse. Repeat.

Once you've replaced *make.c* with your modified version of my code, you're ready to compile it and upload it to your MAKE Controller, just like you did earlier. The filename will still be *heavy.bin*, but it will be in the *WeatherReader* subdirectory. Make sure you've got the power connections in place, plug the MAKE Controller into your network using its Ethernet port, and power it up. If all goes well, you'll see your weather forecast. Otherwise, look for me in the IRC channel and I'll do my best to help!

Brian Jepson is a contributing editor for MAKE, and the co-founder of Providence Geeks, which holds monthly geek dinners at AS220 in downtown Providence, R.I.

BALL OF SOUND

Construct a low-cost spherical speaker array. By Michael F. Zbyszynski

Acoustic instruments radiate sound in a wonderfully complex, 360° fashion, while conventional loudspeakers project a boring spotlight of sound. You can spend a ton of money on a fancy spherical audio array, or you can build one for cheap out of 2 IKEA salad bowls and 8 surplus car speakers.

1. Cut the hole for the top speaker.

I would love to say that I was smart enough to have found a speaker and bowl combination that worked out like this, but it was luck. The speaker fit exactly inside the rubber ring on the bottom of each bowl, after I cut along its inside edge (Figure A).

2. Make a template.

Since the hole for the top speaker was the perfect size, I used it as the template for the other 3 holes.

Trace it onto a piece of paper, and use scissors to cut out the template (Figure B).

3. Mark the side speaker holes.

Use your paper template to trace the other speaker holes onto each bowl (Figure C). The holes should be cut evenly around the sides of the salad bowl. To make sure they are exactly 120° apart, use a piece of heat-shrink tubing (or the like) to measure the bowl's circumference. Then divide the circumference by 3, cut or mark the tubing to that length, and use it to mark where the center of each speaker should be.

For one hemisphere, I centered a speaker on the handle and measured from there. For the other one, I started by centering a speaker on the spout. That way, the speakers are offset when you put the hemispheres together. Make measurements from

Photography by Michael F. Zbyszynski

MATERIALS

4" loudspeakers (8) I would have used the $6, 8Ω speakers from All Electronics (makezine.com/go/speaker1), but they were out of stock. Instead, I bought the 40W, 8Ω shielded woofers (makezine.com/go/speaker2) for $8 each.

The important considerations for the speakers are an impedance of 8Ω (which is normal for home stereo speakers) and good frequency range (in this case 70Hz–10KHz). Sound localization is more acute at higher frequencies, so response >1KHz is especially important. It would be better to add a subwoofer to make up for thin bass than to have no high end.

Quad spring-action speaker terminals (4) makezine.com/go/terminals
Ikea Reda bowls (2 sets) $5 each; use the large bowl in the set of 3 makezine.com/go/ikeabowl
Machine bolts (32)
Locking nuts (32)
Washers (32)
Small nuts and bolts (8) for the terminals. Most of the speakers were fine with ½" bolts, but the top and bottom ones needed longer (1½") ones, as you'll see later. This may vary, if your parts are different.
Foam weatherstripping
Speaker wire I had some 18-gauge stuff lying around the house; the project needs only a few feet.

TOOLS

High-speed rotary tool with #409 cutoff wheel and #561 cutting bit
Soldering iron and solder
Heat gun and heat-shrink tubing
Wire cutters and strippers
Scissors
Pencil
Safety glasses
Earplugs It wouldn't do to deafen yourself in the process of building a snazzy speaker array, would it?

both directions to ensure there aren't any errors.

I decided to put the speakers 1" away from the rim of the bowl. That left enough clearance to put the lids back on. Don't put them any farther away, or they will whack into the top speaker inside the bowl.

4. Cut the side speaker holes.

With the Dremel cutting bit, cut along the lines you just drew (Figure D). There is quite a bit of inaccuracy introduced by drawing a flat template onto a curved surface, so the first hole you cut will be too small. Test fit the speaker, and cut out more where it rubs against the edge. Don't worry about ragged edges; the speaker flange will cover them.

5. Modify the terminals.

The speaker terminals have some extra plastic that prevents them from mounting flush to the bowl. Cut it off using the cutoff wheel.

6. Prepare your speaker wire.

Cut a short (6"–8") length of speaker wire for each speaker. Strip the insulation about ½" back on each end of the wires. Tin all of the ends (Figure E).

7. Solder the wires to the speakers.

It's easier to solder the wires before attaching the speakers to the bowl. Cut short lengths (¾") of heat-shrink tubing and thread them on one end of a speaker wire. Solder the ends to the speaker terminals, slide the tubing over the solder joints, and shrink the tubing with a heat gun (Figure F).

Be sure to maintain consistent polarity. My speaker cable has one reddish and one silver wire, which I connect to the positive (+) and negative (-) terminals, respectively. Other wire has colored insulation, or just a stripe on one side. Conventionally, a red or striped wire is connected to the positive terminal. If you are inconsistent, the speakers will be out of phase, which will probably have no audible effect with this system, but would be a problem with a more scientifically calibrated setup.

8. Prepare the speakers with weatherstripping (optional).

I ran a line of open foam weatherstripping around the edge of all of the side speakers. (The flange of the top speaker already sits flush against the "foot" of the bowl.) The purpose is to fill the gap between the speaker flange and the bowl, and hide any messy edges.

9. Attach the speakers to the bowl.

I started this process with the top speaker. Fit the speaker into the hole, and use a pencil to mark where the bolts should go. Remove the speaker and drill holes that are the same size as your bolts. Put a washer and a locking nut on the back (see Figure G). I used locking nuts because there will be a lot of vibration.

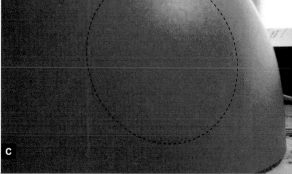

Fig. A: Hole cut in the bottom of the bowl.
Fig. B: Template traced from hole piece.
Fig. C: Side speaker position traced onto the bowl.

Fig. D. Placement of side speakers.
Fig. E. Speaker wire ends, tinned with solder.
Fig. F. Speaker wire soldered to the speakers.

Fig. G: Speaker bolted in with washers and locking nuts.
Fig. H: Terminal connected to the side of the bowl.

Fig. I: Two different terminal placement options: on opposite sides of the speaker, or together near the handle. Fig. J: Terminals connected, speakers all wired.

There's a fair bit of muscle and bending in this step. Both the speakers and the bowl might become a bit bent by the end. That's fine; it just makes the whole thing stronger.

Repeat for the 3 side speakers.

If you accidentally dent a speaker, check out this Instructable on "unpopping" dented speakers: makezine.com/go/unpop.

10. Attach the terminals.

The terminals each need a hole big enough for the connections to fit through. Hold the terminal in the approximate location, and mark the place to cut. The terminal block is much bigger than the necessary hole, so a little imprecision is fine (Figure H).

I was imagining this speaker hanging from the handles, so I kept the terminals as close to them as seemed reasonable. Check out Figure I to see how the positions are different on each hemisphere. There are 2 terminal blocks on each hemisphere. On one, they're close together near the handle. On the other, they flank the speaker that is centered on the handle.

Stick some weatherstripping to the bottom of the terminal blocks, then fit them onto the bowls, drill holes for the bolts, and bolt the blocks in place.

11. Connect the terminals.

Thread a piece of heat-shrink tubing onto each speaker wire, and solder the pairs to the terminals, making sure to solder the positive wire (red, striped, etc.) to the positive (red) terminal.

After the joints cool, slide the tubing over them and use your heat gun to shrink the tubing (Figure J).

12. Enjoy the sweet, sweet music.

This array can be used either as a hanging spheroid, or as 2 hemispheres. Either way, the radiation pattern can be quite interesting. I like to put the lids on when they're in hemisphere mode.

Thanks to Dan Truman and the researchers at CNMAT, whose scientific work inspired this project.

Michael Ferriell Zbyszynski is a composer, sound artist, performer, and teacher in the field of contemporary electroacoustic music. Currently, he is assistant director of music composition and pedagogy at UC Berkeley's Center for New Music and Audio Technologies (CNMAT).

Make:
technology on your time™

Sign up now to receive a full year of **MAKE** (four quarterly issues) for just $34.95!*
Save over 40% off the newsstand price.

_____
NAME

_____
ADDRESS

_____
CITY STATE ZIP

_____
E-MAIL ADDRESS

MAKE will only use your e-mail address to contact you regarding MAKE and other O'Reilly Media products and services. You may opt out at any time.

*$34.95 includes US delivery. For Canada please add $5, for all other countries add $15.

makezine.com/subscribe
For faster service, subscribe online

promo code **B79TB**

makezine.com

Make:
technology on your time™

Give the gift of MAKE!

makezine.com/gift
use promo code **47GIFT**

When you order today, we'll send your favorite Maker a full year of MAKE (4 issues) and a card announcing your gift—all for only $34.95!*

Gift from:

_____
Name

_____
Address

_____
City   State

_____
Zip/Postal Code   Country

_____
email address

Gift for:

_____
Name

_____
Address

_____
City   State

_____
Zip/Postal Code   Country

_____
email address

*$34.95 includes US delivery. Please add $5 for Canada and $15 for all other countries.

47GIFT

BUSINESS REPLY MAIL

FIRST-CLASS MAIL   PERMIT NO 865   NORTH HOLLYWOOD CA

POSTAGE WILL BE PAID BY ADDRESSEE

Make:

PO BOX 17046
NORTH HOLLYWOOD CA 91615-9588

BUSINESS REPLY MAIL

FIRST-CLASS MAIL   PERMIT NO 865   NORTH HOLLYWOOD CA

POSTAGE WILL BE PAID BY ADDRESSEE

Make:

PO BOX 17046
NORTH HOLLYWOOD CA 91615-9588

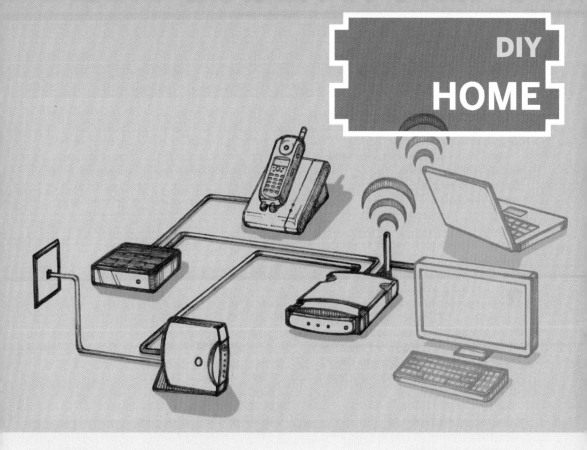

FREE VoIP

Got broadband? Add phone service for $0/month. By Dave Mathews

These days, many people don't even have landline-based telephone service; they get all their calls on their mobile and office phones. But once I settle in at home, I want people to be able to call me on a nice, comfortable cordless phone, even if it's lost in the couch. At home, the cordless battery is always charged, and it doesn't matter if I'm in a fringe mobile service area.

If you already have a broadband connection, you can get a home phone without paying for POTS (plain old telephone service) or a Vonage-style digital phone service. Here's how to do it the easier way, without having to run your own private branch exchange (PBX) Asterisk server.

Roll Your Own

To roll our own free VoIP, we're going to configure

some telephone adapter hardware, sign up for some free internet services, and reconfigure our broadband router. It's not the simplest arrangement, and routers can be unforgiving, so follow these directions closely.

1. Acquire the phone hardware.

You can use either a SIP (session initiation protocol) VoIP phone from Pingtel, Linksys, or Polycom, or an analog telephone adapter (ATA) from Linksys/Sipura or Cisco. Make sure the device is not locked to a provider. Adapters from Vonage are subsidized and therefore cheaper, but won't work with outside applications unless someone unlocks them. Expect to spend about $50 on eBay for an unlocked ATA. I chose the Sipura SPA-2000 ATA, which has 2 phone jacks on one side and an Ethernet jack on the other.

ROLL-YOUR-OWN VoIP CANS AND CANNOTS

Your phone will be able to:

- Receive incoming calls from any phone, mobile, landline, or VoIP.
- Dial many other VoIP users, but not all of them — see below.
- Dial toll-free numbers in the U.S., U.K., Netherlands, Norway, and Germany, and use calling cards for those countries to reach any number.
- Call any Free World Dialup (FWD) members using just 6 digits on your keypad.
- Dial any number at all by going through a commercial SIP service such as Broad-Voice, VoicePulse, or VoipBuster, instead of FWD. BroadVoice and VoicePulse charge monthly, and VoipBuster charges per minute — less than 10¢/minute for most countries.

Your phone will not be able to:

- Dial land or mobile phones without SIP service or a calling card.
- Dial Vonage or Skype users.
- Use a Vonage hardware-based phone adapter (but you can use their SoftPhone software app within your hardware adapter).
- Dial 911. For this, use a mobile or landline. Even inactive mobile phones support emergency dialing.

2. Sign up for an FWD account.

FWD is a large PBX run by Jeff Pulver with a pretty good up-time record. To get started, browse to freeworlddialup.com, click on the my.FWD tab, then click Sign Up for Fwd on the left, and fill in your details. Don't sign up for voicemail; we'll set that up later elsewhere. Write down the 6-digit phone number you're assigned; this is the number you'll share with your friends on FWD and use for configuration.

3. Get a U.S. number from IPKall.

They are nice enough to provide anyone with a Washington state phone number, which can then route to your FWD account. Head to ipkall.com, click Sign-Up, and choose any area code that you want, as long as it's in Washington. Then enter your 6-digit FWD number from the previous step. Keep the

proxy set to *fwd.pulver.com* and enter your email address. Choose a 4-digit password, and enable voice-mail as well. I chose a 40-second ring until voicemail answers and emails you an MP3 of the message.

Note that the free IPKall service takes an hour to activate, and if you don't use the number you're issued for a period of months, it will be assigned to someone else. Thanks to bad acts of Congress, IPKall can make all their money off termination fees from other exchange carriers.

4. Reconfigure the firewall.

Now comes the tricky part. First, choose a fixed IP address for your ATA or VoIP hardware or for the PC that's running the VoIP software. This address should never change. Next, log in to your broadband router configuration screens, open TCP or UDP port forwarding, and add rules to pass all external traffic to that IP address for the following ports:

- » **5060:** TCP/UDP (for signaling traffic)
- » **5082:** TCP/UDP (for NAT traversal)
- » **16384–16482:** UDP (for UDP voice sessions)
- » **4569:** TCP/UDP (IAX2 port for IPKall inbound)

This allows call setup, ringing, and the all-important voice traffic. Different routers put these settings under different headings, such as Application, External Port, or Virtual Server Forwarding, which is not nice.

5. Configure your SIP device.

Now it's time to configure your VoIP phone or ATA. Follow the manufacturer's instructions to enter your chosen fixed IP address into the configuration page or screen of your phone or adapter, and log in. Supura devices like the SPA-2000 require you to log in as " Admin" (at the bottom left) in order to change settings.

Here are the settings to use:

- » **Outbound Proxy:** fwdnat.pulver.com:5082
- » **Proxy:** fwd.pulver.com
- » **SIP Port:** 5060
  NOTE: Some configuration fields have a unique area for the port setting (e.g. 5060 or 5082) and others require it after the domain.
- » **Register:** Yes
- » **Display Name:** Your IPKall or FWD phone number — your caller ID
- » **User ID:** Your FWD number
- » **Password:** Your FWD password
- » **Codec:** G711u
  NOTE: Use this for now; you can switch to a lower bit rate later.

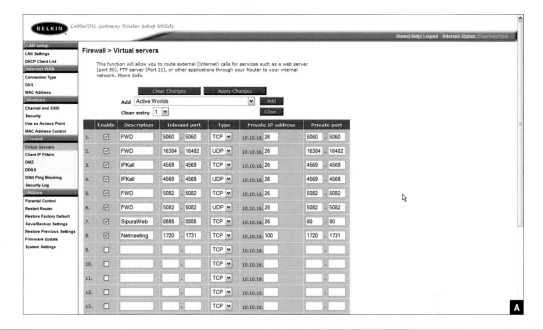

Fig. A: Configuring your router's firewall to enable VoIP by letting through data packets for call setup, ringing, and voice.

» Dial Plan: (xxx|xxxx|xxxxx|xxxxxx|*3xxxx|*4xxxx|*1xxxxxxxxxx|**xxxxxxxxxx)
NOTE: Your device may not support this setting, which defines valid keypad inputs for dialing.

Since you're probably behind a firewall, you also need to point it to a STUN server in order to direct calls around it. (STUN is the handy acronym for "Simple Traversal of UDP [User Datagram Protocol] through NATs [Network Address Translators].") With a Sipura adapter, you do this by enabling STUN and all of the VIA options except for Insert VIA rport. Use the STUN server *stun.fwdnet.net:3478* and disable Send Resp to Src Port and STUN Test Enable.

All-soft alternative: If you don't need a traditional phone or RJ-11 connectivity, you can set up free VoIP with any SIP VoIP phone or even an Asterisk switch. For a software-only version, download free SJphone PC software from sjlabs.com/sjp.html and talk on a headset. One version is preconfigured for FWD, so you can go the easy route and skip Step 5.

6. Make some test calls.
First, dial 613 and see if you can hear the Asterisk operator. Speak, and you should hear yourself a fraction of a second later. The longer the delay, the more latency on your connection and the more echo your callers will hear. Ideally, you'll hear yourself almost immediately.

You can also check the time or your phone number by dialing 612 and 958, respectively. See "VoIP Phone Tricks" on the following page for more calls you can make.

7. Call your new home number.
If it's been at least an hour since you got your IPKall Washington state number, dial it from your mobile phone. Your VoIP line should ring. Now you can hand that number out to your mom, or to people whom you do not want using up your phone minutes, like credit card companies or your utilities.

Troubleshooting
If your adapter cannot register, or you can make calls but don't hear anything, then you have a NAT or firewall issue. Try reflashing your firewall's firmware or reconfiguring. If you hear a ring, but then hear a fast busy signal when the party answers, you have a codec mismatch between adapters. You might try uLaw protocol on your adapter, although G711u is usually the most compatible. Advanced users can try G729, which is a low bitrate codec.

Fig. B: Configuring a Sipura ATA (analog telephone adapter) to work with your Free World Dialup (FWD) account and your free Washington state phone number from IPKall.

VoIP Phone Tricks

To dial a toll-free number in the United States, enter *1 followed by the number. For example, the Frito-Lay Customer Service number would be *18003524477 (*1-800-FL-CHIPS).

Note that some toll-free numbers route to different call centers based upon your location, and may not work correctly for you because they think your number is coming from another physical location. Some call systems may even just hang up on you.

For other countries, try the following: Netherlands *31800; U.K. *44800, *44500, *44808; Germany *49800, *49130. These will let you use local calling cards for the countries, but some calls may not connect due to peering issues. There's also a limit to the number of minutes you can talk, so your mileage will vary.

There you have it — open toll-free calling and inbound calls into a device much like what Vonage provides its customers. Sure, the services are more limited (and you may have killed yourself after that firewall configuration), but you get what you pay for! Remember that IPKall numbers are "use it or lose it," since they recycle inactive numbers. In fact, you may lose it anyway; there are no guarantees.

But in the meantime, as you get all your friends to sign up for FWD, here are some numbers to call and keep you busy:

411 — Directory information through the voice portal TellMe
612 — Time
613 — FWD Echo Tester: you talk and should hear yourself a millisecond later.
958 — Phone number confirmation
*18xx — U.S. toll-free (don't forget the asterisk!)
55555 — Talk to the FWD volunteer staff

If you need a few minutes of outbound calling to mobile or home phones, check out the free VoIP services listed at makezine.com/go/freevoip.

For pay VoIP service providers, see makezine.com/go/payvoip.

Dave Mathews is a writer, inventor, and hardware hacker based in San Francisco. More stories on Asterisk and VoIP can be found on his website at davemathews.com.

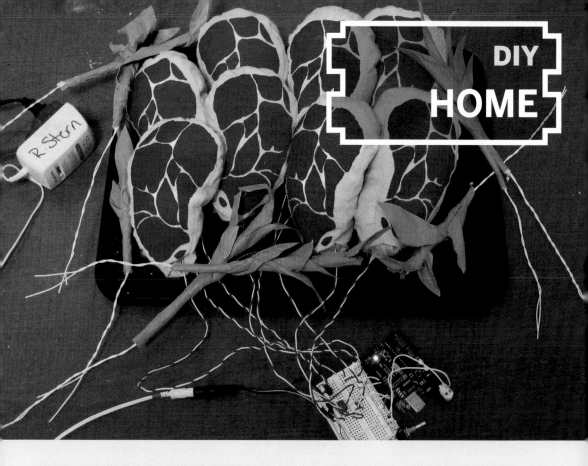

PLUSH IRRADIATED SIRLOIN

Microcontroller night light illuminates meaty issues. By Rebecca Stern

Photography by Rebecca Stern

Faced with an assignment to make a plush night light, I thought, "Why light?" and brainstormed reasons for a stuffed toy to light up. In a glowworm toy, for instance, the light mimics nature. I'd been reading Michael Pollan's *The Omnivore's Dilemma*, and this got me thinking about the chain of refrigeration, labor, and irradiation involved in American beef production. So I thought, glowing irradiated meat! I know that irradiated meat doesn't glow, and neither does toxic waste unless it's in a cartoon, but plush toys typically represent cartoon characters anyway, so it made sense: Plush Irradiated Sirloin.

1. Prepare the fabric.

I silk-screened my steak illustration onto pink flannel and sewed the pieces together (inside out, so the seams wouldn't show), leaving a small opening at the base of each one. (*For an excellent primer on silk-screening, check out CRAFT magazine, Volume 01, page 106.*) You can also use pre-patterned fabric or use fabric paint to hand-paint the design. Next, I turned them right side out, but left them empty. I had to put the lights inside before I stuffed the plush fiberfill around them!

2. Add the Arduino board.

Inside, each steak has two 360° super-bright LEDs wired in series. These have frosted lenses that distribute the light evenly in all directions, making them perfect for the inside of plush toys. Because I wanted the steaks to glow dimmer and brighter periodically, I needed some kind of signal to control the brightness of the lights. The Arduino board, my favorite microcontroller solution lately, supports

MATERIALS

360° LEDs (16) **available at** superbrightleds.com
TIP 120 transistors (2)
Solderless breadboard **and/or solder-type**
   breadboard
Wire
Solder
Patterned fabric **or silk-screen your own**
Thread
Polyester fiberfill plush stuffing
Arduino board with power connector
12V or adjustable AC power adapter
Toggle switch **(optional)**
Epoxy **or hot glue**

TOOLS

Soldering iron
Iron and ironing board
Sewing machine
Hand-sewing needle and sewing pins
Pliers
Wire cutters/strippers
Hot glue gun **if using hot glue**
Computer with Arduino IDE
USB A-B cable

The breadboard and Arduino.

the perfect feature for this: pulse width modulation (PWM). PWM can make LEDs, which are binary, appear dimmer by pulsing them on and off, with varying time ratios, faster than the human eye can detect. I could use this to produce the analog sine-wave-like throbbing glow that I wanted.

The PWM signal controls the glow, but the Arduino can only output up to 5V, which isn't high enough to power these super-bright LEDs. I had planned to power the Arduino with a 12V AC adapter, so

I designed the circuit to drive the LEDs from the same source. I used 2 TIP 120 transistors to amplify the signal to each half of the meat tray, 4 steaks each. This pumps the circuit's full 12V through 2 parallel sets of 2 LEDs (2 steaks, 4 LEDs) in series, which works out to 3V per LED.

3. Add the LEDs.

For each steak, I made an LED insert with 2 LEDs wired in series and neatly twisted. I spaced the LEDs about 4" apart, so that they would each light up an even half of the steak without being too close to the edges. I made the lead wires really long, and I knew they would be exposed, so I chose red and white wire to match my plush.

After wiring up the circuit and soldering and testing the LEDs, I finally assembled the steaks. It's important to make sure all your LEDs are functioning properly first; it's no fun to debug a sewn-together toy. Since electronics with fabrics could be a fire hazard, I covered the LED leads in epoxy (hot glue works, too) to prevent a potentially dangerous short.

I positioned each double-LED wire inside a steak, and filled around it with polyester filling. I left the LEDs plugged in, so I could see how the light diffused and adjust them accordingly. When I got them how I wanted, I stitched up the bottom openings by hand, and arranged them together on a tray.

4. Bask in the glow.

Each half of the tray (4 steaks) glows in alternation with the other. The pattern is subtle and soothing, the way a good nightlight should be. They're soft, but not very cuddly, as they remain tethered to their circuit board. In the future I could embed smaller circuit boards inside each steak to make a portable, more snuggly version. I've also been thinking of making a larger version for throw pillows, or a smaller version with catnip instead of electronics. These steaks have been great conversation starters in the classroom and online, and I hope they inspire people to learn about the politics of our food industry.

➕ For a full schematic of the circuit, the microcontroller code, and the pattern for the steak silkscreen, visit makezine.com/11/diyhome_steak.

Rebecca Stern of Brooklyn, N.Y., is an artist, technologist, and recent graduate from Parsons School of Design, where she studied design and technology. sternlab.org

ON BATTERIES

How to decide which batteries will run your project best. By Limor Fried

Photography by Sam Murphy

If knowledge is power, then knowledge of batteries is power squared! Here's what I know about disposable and rechargeable batteries and their tradeoffs.

Measuring Batteries

There are a few ways you can measure batteries of different types. Here's what I'll be comparing:

Size There are standard sizes such as AA and 9V, an assortment of coin cells, and special rechargeable power modules for specific products.

Voltage Labeled voltage and actual voltage differ. An alkaline labeled "1.5V" actually starts out at 1.6V, quickly drops down to 1.5, and then drifts down tery's voltage derives from its chemistry. Alkalines are always 1.5V, lead-acid cells are 2V, and lithium

cells are 3V. Batteries with higher voltages like 12V contain multiple connected cells in one package.

Reusability Rechargeable batteries vary in the number of recharge cycles possible during their useful life, usually in the hundreds. But you need to use a high-quality charger, one with sensors; cheap chargers can kill your cells. NiCd and NiMH rechargeables have lower voltages than alkalines, so test them before using them as replacements.

Power capacity determines how much power a battery holds. On the packaging, this is listed in milliampere-hours (mAh), a measure of charge capacity. To convert this to power capacity (watt-hours or Wh), multiply by the voltage.

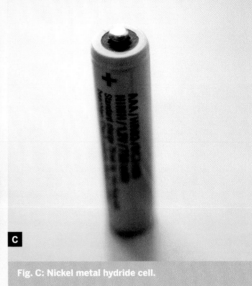

Fig. A: Lead-acid battery. Fig. B: Lithium-ion battery.

Fig. C: Nickel metal hydride cell.

Power capability is the rate at which a battery can deliver its power without wasting energy, denoted by C and measured in amps. The C of a battery is proportional to the capacity, so a 1Ah capacity battery with 0.1C capability can supply 100mA of current. Stacked batteries should have matching C values, because total capability is limited by the weakest link.

Capability depends on chemistry. Lead-acid batteries have the highest capability, 10C or more, which is why they're used for cars and other applications that require floods of power. Alkalines are about 0.1C, and lithium cells have capabilities of only about 0.01C — just enough trickle to run a wristwatch.

Power density refers to how much power a battery supplies for its weight, expressed in watt-hours per kilogram (Wh/kg). Higher-density batteries are good for projects that need to be lightweight.

Price tends to be proportional to capacity and power density. The more power you want in a smaller, lighter package, the more you have to pay.

Battery Chemistry Types

Lead-Acid
These workhorses are good for systems that need plenty of power where weight is not important, such as big motors, projectors, and loud amplification (Figure A). Lead-acids are based on 2V cells, with the most common voltages being 2V, 6V, 12V, and 24V.

Pros: Rechargeable, cheap, powerful, high capability
Cons: Very heavy, batteries tend to be large because power density is low
Price and capacity: $20 for a 12V battery with 7Ah
Power density: 7Wh/kg

Alkaline
These common disposables are great for projects that need to be user serviceable (Figure D). Cells are 1.5V, available in sizes from coin cells to AAAA to D, so it's easy to up-size to get more capacity.

Standard 6V lantern batteries (Figure E) and 9V batteries (opening photo) are multi-cell alkalines. Lantern batteries have massive capacity and terminals that are easy to clip or solder to. Standard 9V batteries are common but have low capacity and capability, are expensive, and can't continuously deliver more than 20mA.

Pros: Popular, safe, long shelf life
Cons: Non-rechargeable, low capability
Price and capacity: $1 for a AA cell with 3,000mAh
Power density: 100Wh/kg

NiCd (Nickel-Cadmium)
The original rechargeables, "ni-cads" are cheaper than NiMH and are still used in cordless phones and other products. NiCds also hold their charge longer than NiMHs when not in use. Cells are 1.2V, often bundled into 3.6V batteries (not pictured).

Pros: Rechargeable, inexpensive, long-lasting, standard sizes

Fig. D: Alkaline cell.

Fig. E: 6V alkaline lantern battery.

Fig. F: Lithium coin cell.
Fig. G: Lithium battery.

Cons: Low power density, require periodic full discharge/recharge cycles to reduce "memory" (caused by crystals growing on internal plates), contain toxic cadmium metal
Price and capacity: $1 for a AA cell with 1,000mAh
Power density: 60Wh/kg

NiMH (Nickel Metal Hydride)
With cell voltage of 1.25V, these rechargeables (Figure C) are a good replacement for alkalines. They like to be charged at about 0.1C but can be discharged at 0.2C.
Pros: Rechargeable, high power density, standard sizes, higher capability than alkalines
Cons: More expensive and shorter service life than NiCd, self-discharge quickly
Price and capacity: $2 for a AA cell with 2,500mAh
Power density: 100Wh/kg

Li-ion (Lithium-Ion) and Li-poly (Lithium Polymer)
These are the latest rechargeables, and are now standard for portable consumer electronics (Figure B). They are very light and have high power capability and density, but also require special circuitry to keep them from exploding! That's why you can't buy general-purpose Li-ion cells in standard sizes.

To use these, your best bet is a camcorder (large), cellphone (medium), or R/C airplane (small) battery with its matching charger. Cells are 3.6V, commonly packaged as 3.6V and 7.2V. They can easily provide 1C of current, and some deliver 10C!

Pros: Rechargeable, ultra-light, high cell voltage, high capability, high density
Cons: Expensive, delicate, can explode if misused!
Price and capacity: $10 for a replacement cellphone battery with ~750mAh
Power density: 126Wh/kg for lithium-ion, 185Wh/kg for lithium polymer

Lithium Batteries and Coin Cells
Lithium batteries have a voltage of 3V per cell (Figure G). Most are in coin/button form (as in Figure F, but coin cells can also be alkaline, or zinc-air "hearing aid batteries" — all 1.5V). These are great for small, low-power devices, but they can't be recharged and provide only 0.005C of continuous current (although you can draw more in pulses).

One popular cell is the CR2032, which measures 20×3.2mm and provides 220mAh. Coin cells can get as large as the 24×8mm CR2477, and the 3V lithium CR123 is a bit thicker and shorter than a AA.
Pros: Light, small, inexpensive, high cell voltage, high density, easy to stack, long shelf life
Cons: Non-rechargeable, low capability, need a special holder
Price and capacity: $0.35 for a CR2032 with 220mAh; $1.50 for a CR123 with 1,300Ah
Power density: 270Wh/kg

Limor Fried makes nifty things and writes about them on ladyada.net.

ELECTRONIC CRICKETS

Create a nighttime chorus by modifying solar yard lamps. By Michael F. Zbyszynski

Much of my creative work as a composer and sound artist relies on multimedia computers to generate and manipulate sound. Unfortunately, this method has a limitation: I need plenty of electricity, preventing me from installing art in parks and outdoor spaces without AC power.

Rather than ignore such opportunities, I started thinking about solar power. It was David Zicarelli (of Max/MSP fame, see cycling74.com) who suggested I look at the "solar garden lamps" that are ubiquitously available in hardware and home stores.

For a maker, solar lamps represent a tremendous bargain: for $5 you can get a solar cell, batteries, charging circuitry controlled by a photoresistor, and a functional, waterproof housing. I've never been impressed with these in their intended function as a light source, but they are an interesting platform

for more creative use.

For this project, I wanted to make something that reminded me of many beautiful phenomena of summer nights: crickets, chirping frogs, and fireflies. By day, the lamps seem ordinary; they sit and charge their batteries like all their unaltered cousins. But as the sun goes down, each one starts blinking and chirping. The sound and rate of their song are determined by the temperature, the amount of sun they receive, and the natural variance of their components. The emergent quality of dozens together can be fascinating. This project has a certain affinity to BEAM robotics (*see MAKE, Volume 06, page 76*).

It took some shopping to find the right lamps. I wanted to use a large number (my current installation has 20), so price was a major factor. Also,

A

Fig. A: Half of this circuit is a modified astable multivibrator, a handy circuit for blinking things. I slipped in a thermistor in place of one of the ordinary resistors, so that the timing would be temperature-dependent.

The other half of the circuit is a Hartley oscillator. I tuned the component values up so that it sounded the way I liked at the relatively low voltage I was expecting.

MATERIALS

Solar lamp
Small piece of stripboard aka Veroboard
Xicon Ultra-Mini Transformer 500 CT-8
  (Mouser part #42TL001-RC)
Speaker 1" diameter with polypropylene cone
10Ω resistor and 100Ω resistor
4.7K resistor and 15K resistor
20K NTC thermistor
200K potentiometer, single turn
0.01µF ceramic capacitor
0.022µF ceramic capacitor
6.8µF electrolytic capacitor
100µF electrolytic capacitors (3)
Transistors (3) PN2222A or any small signal NPN
LED Green, water clear, standard output
Wire I used solid core, but 22 AWG stranded would
  be perfect.
Screwdriver(s)
Straightedge
Multimeter
Hot glue gun and glue

SOLDERING KIT

Soldering iron and solder
Desoldering braid and/or pump
Third hand
Lead trimmers
Small pliers

typical garden lamps come with either 1 or 2 AA batteries as a power source. Two batteries in series yields double the voltage (nominally about 2.4 volts for rechargeables), so I preferred those lamps. I found a 4-pack of lamps at my local mega home store for $20. I think stores discount these heavily on occasion, as a loss leader, so look around a bit.

In designing the electronics for this project, I decided early on not to use any microcontrollers. I wanted everything to be as simple and generic as possible. My goal was to make a circuit that blinked an LED every second or so, depending on the temperature, and simultaneously triggered some sound generator that I found pleasing. The final design looked like Figure A.

1. Make the circuit.

Refer to Figures A and B to build the circuit.

2. Open up the lamp and remove the LED.

Unscrew the housing to find a circuit board, also screwed in. Carefully unscrew it, leaving it connected to the solar cell and photoresistor, and use a soldering iron and desoldering pump (or braid) to remove the LED from the circuit board. Save this

B

Fig. B: I was given a bunch of old Veroboard by my colleague Adrian Freed (one of his many contributions to this project). Snap the board to the right size using a straightedge. Solder the parts in as per the schematic. I tried to keep it fairly compact, so it would fit inside the lamp housing. I cut off the left end of the stripboard (including the LED, which is there for testing) before installation.

LED for another project. Remember the location of the LED because that's where you'll add the custom circuit in the next step.

3. Install the circuit.

Use short wires to connect the positive (check with a multimeter) and negative pads, where the old LED was, to the power and ground of the custom circuit.

Next, attach a new LED and speaker to your custom circuit. I ran the wires for the speaker through a hole in the bottom half of the housing, and used the speaker magnet to stick to a battery. Different speaker positions create different sounds, so feel free to experiment. Use a shorter pair of wires to attach the new LED to the custom circuit. The new LED probably will not fit perfectly into the space of the old LED, so liberally hot-glue it. The hot glue also provides insulation and weatherproofing. Do not cover the lens of the LED with glue.

4. Tune up your lamp.

Cover the photoresistor with your hand or a piece of tape, and see how it sounds (make sure the batteries are charged). Adjust the potentiometer to get a good sound, making sure that each lamp has its own tone that is different from its neighbors.

5. Close up the lamp.

Delicately bend the extra wire and the 2 circuit boards into the housing of the lamp. Go slowly, and it shouldn't be too hard to fit everything in and screw the lamp shut.

6. Install and enjoy.

These lamps have the same needs as their standard brethren. Put them someplace that gets a fair amount of sun and is out of the way of foot traffic. The most interesting effect is when you can see and hear a number of them. They are quite visible, but fairly quiet. I have 20 lamps encircling a medium-sized yard, with about 2' between each lamp.

As the twilight begins they will each start chirping, tentatively at first. By sunset they will all be singing.

See the cricket lamp in action:
makezine.com/11/diyoutdoors_cricket

Michael Ferriell Zbyszynski (mikezed.com) is a composer, sound artist, performer, and teacher in the field of contemporary electroacoustic music. He is assistant director of music composition and pedagogy at UC Berkeley's Center for New Music and Audio Technologies.

DIY ECGs

TRACK YOUR TICKER WITH A HOMEMADE ELECTROCARDIOGRAM MACHINE.

By Dr. Shawn

THE HEART'S PUMPING ACTION IS DRIVEN by powerful waves of electrical activity that cause weak currents to flow in the body, changing the electric potential between different points on the skin by about one thousandth of a volt (one millivolt, 1mV). Hidden within that activity is an enormous amount of information about what the heart is doing, and anyone who can detect it can peer into the workings of this incredible organ.

Fortunately, you don't have to be a cardiologist with expensive equipment to pick up and decipher that signal. Anyone can do it with this homemade electrocardiogram (ECG) device, an analog-to-digital converter (ADC) to digitize the signal and send it to a computer, and a remarkable book that I'll tell you about later.

You can assemble the circuit itself in an afternoon for about $40. The ADC will cost a bit more, between $50 and $300. But these devices open a universe of opportunities to the home-based experimenter, and so every citizen scientist should invest in one. (I've negotiated a great deal on one of these devices especially for MAKE readers. Read on.)

The experimental challenge is that the signal we're looking for measures only about 1mV, it can change in as little as $1/100$ of a second, and it's embedded in a noisy environment. To keep up with the signal and boost it to a digitizable 1V level, you need an amplifier with a gain of about 1,000 and a frequency response of at least 100Hz. But standard operational amplifiers (op-amps) like RadioShack's 741 won't work because of the surrounding noise.

When electrodes are placed far apart on the body, our skin acts like a crude battery and generates an irregular potential difference that can exceed 2V, dwarfing our 1mV heart signal. Even worse, your body and the wires that connect to the electrodes make wonderful radio antennas that pick up the 60Hz hum emanating from every power cable in your home. This adds a sinusoidal voltage, which further swamps the tiny pulses from your heart, and because its frequency lies close to the 100Hz resolution we need to track your heart, it's hard to filter out.

Now, you electronics types might think this shouldn't matter because op-amps are "difference amplifiers" — that is, they subtract out any voltage that runs equally to both inputs. Unfortunately, op-amps don't do that job perfectly, and when the swells are thousands of times bigger than the signal, as they are here, you're sunk. To ensure that this "common-mode" garbage adds no more than a 1% error to our measurement, we need what's called a common-mode rejection ratio (CMRR) of at least 100,000 to 1. In electronics parlance, CMRR is measured in decibels (dB), where every factor of 10 increase in voltage is equivalent to 20dB. This makes our required ratio 10^5, which equals 20*5 or 100dB — a precision beyond that of most op-amps.

THE INSTRUMENTATION AMPLIFIER

When an application calls for both high gain and a CMRR of 80dB or greater, experienced experimenters often turn to special devices called instrumentation amplifiers. These remarkable devices were once bulky and expensive, but today, they can be purchased for just a few dollars as an integrated circuit. To make the construction as simple as possible, I designed this ECG around the Rolls-Royce of instrumentation amplifiers, the AD624AD from Analog Devices, which you can buy from Digi-Key (digikey.com) for under $25. You select a gain of 1,000 with the AD624AD by simply shorting certain pins together, and at this setting the amp's CMRR exceeds 110dB.

The AD624AD is easy to use, but experienced gadgeteers should also feel free to experiment with less expensive options, such as the Analog Devices AD620AN. And if you're a real Daniel Boone-type maker, you can construct your own instrumentation

Citizen Scientist

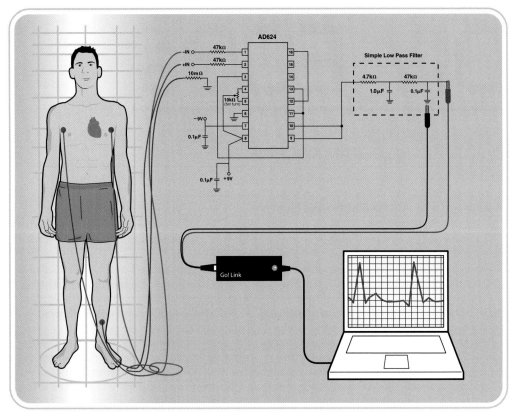

Schematic for a simple instrument to monitor one's electrocardiogram (easily made at home).

amplifier using three RadioShack op-amps and a handful of 100K resistors. (See the complete schematic on the next page.)

Some simple circuitry supports our instrumentation amplifier. A two-stage resistor-capacitor (RC) filter weeds out frequencies higher than about 50Hz. As filters go, this one is pretty wimpy, but it works well enough to do the job. I used a four-wire phone cord to carry the signals between my body and the amplifier, but you only need three of the wires. The side of my project box sports a phone jack for easy connection and disconnection.

THE ELECTRODES

I fashioned my first electrodes out of quarters smeared with a conducting layer of shampoo, taped firmly to my body, and connected to wire leads. They worked, sort of. Then I discovered anyone can buy bags of the real thing — the peel-and-stick electrodes used by cardiologists. The cost is about $13 for 50. (Google "ECG electrodes" for a host of suppliers.) I used alligator clips to connect the signal wires to the metal nipples on the backs of the electrodes.

Connect the instrumentation amp's negative lead to just under your subject's left armpit and its positive lead just under the right armpit. You must also connect a ground lead for the circuit to the left shin just above the ankle. Without the leg as a ground, bad things happen to the signal, and it's a great little experiment to record an ECG this way for about ten minutes and see the problems that creep in.

LOGGING THE DATA

To examine the ECG signal, you'll need to digitize and record it on your computer. This requires an ADC or data logger device that can sample at 200Hz. (The Nyquist Theorem states that reading an oscillating signal requires sampling it at a minimum of 2x its frequency.) I've tried many data loggers, and my favorite is the Go Link from Vernier Software, which has 12 bits of resolution and samples at up to 200Hz. Add the matching voltage probe unit, and you're ready to rock-and-roll literally hundreds of other science projects. I've negotiated a deal with Vernier especially for MAKE readers: $67 for both logger and voltage probe; see sas.org/make.html for details.

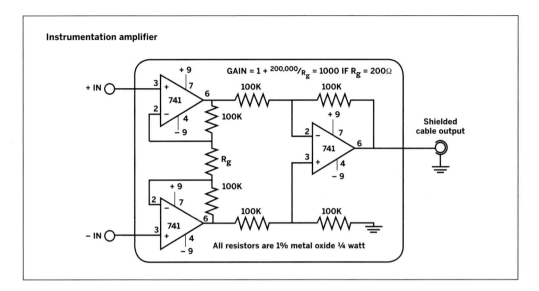

Instrumentation amplifier

$$\text{GAIN} = 1 + \frac{200{,}000}{R_g} = 1000 \text{ IF } R_g = 200\Omega$$

All resistors are 1% metal oxide ¼ watt

Once you've connected body electrodes to our circuit, plug the Go Link into your laptop's USB port, plug the voltage probe into the Go Link, and wire the probe to our circuit's ground and low-pass filter output. For safety, the laptop should be unplugged.

SAFETY

Unless you're doing something exotic, battery-powered instruments are generally safe to connect to people. The Go Link is powered through its USB port, so if the laptop you're using is unplugged, you've got no worries.

Unlike the editors of MAKE, I personally think it's OK to use our ECG device with a plugged-in computer, provided you take some additional precautions; see more discussion on my website at sas.org/make.html. The 10M (that's 10 million ohms!) resistor between the subject and ground will choke any current in the unlikely event that an AC adapter fails and the laptop fries. The 47K resistors also limit the current. Even with a freak power surge, the subject's arms lie outside of any path between the electrodes, so if she experiences any pain, she'll be able to grasp and pull off the wires. Keep someone else in the room, or if that's not possible, make the electrode leads short and run ECGs while standing. That way, if lightning does strike, you'll break the connection by falling down.

In any case, never attach electrodes to the arms, and don't connect the device to anyone in weak health. No matter how much of a do-it-yourselfer you are, don't fire up your homemade ECG if you think you might be having a heart attack. Resist your curiosity and dial 911.

In hospitals, some monitoring equipment adds yet another layer of safety: opto-isolation. In this scheme, battery-powered devices connected to the patient transmit readings via LEDs to matching optical sensors inside a wall-powered display device. Since the only link between the two devices is a stream of photons, large currents can't reach the patient through any wires. If you care to, you can add opto-isolation to your ECG to protect against a meteor-falling-from-the-sky-probability-level chain of events. Opto-isolators come in DIP-style ICs that you can plug into your system between the filter and the data logger. You'd ground the LED transmitter side to the amplifier circuit's ground, and the photodiode receiver to the data logger or to ADC's ground.

INTERPRETING THE RESULTS

Once you've mastered the art of tracking your ticker, you'll need to decipher what it all means. There's just one must-have reference, *Rapid Interpretation of EKG's* by Dale Dubin, MD. You'll find this easy-to-master mainstay of medicine on nearly every doctor's bookshelf in America. Dr. Dubin also happens to be a good friend of the Society for Amateur Scientists, so we're able to offer new copies of this classic to MAKE readers at a 30% discount; see sas.org/make.html.

Dr. Shawn (Shawn Carlson, Ph.D.) is the founder and executive director of the Society for Amateur Scientists (sas.org) and a MacArthur Fellowship winner.

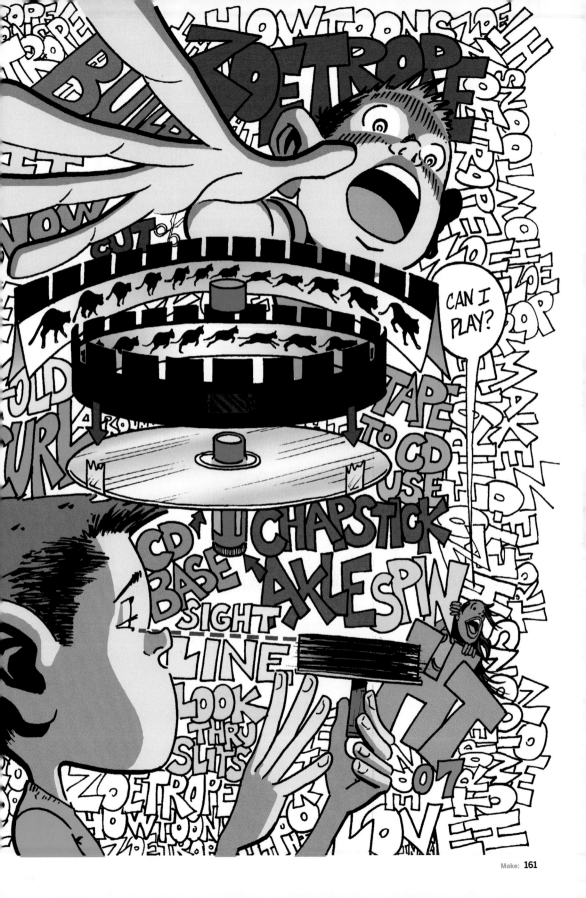

MakeShift

By Lee D. Zlotoff

The Scenario: Having heard the call of the wild, you recently purchased a small, rustic cabin that's situated among tall trees on the edge of an official wilderness area. It has running water and electricity but no landline phone yet. You haven't met any of your neighbors, and the nearest ones are about a quarter-mile away along a dirt road.

You've driven up there for the weekend to install security lighting around the perimeter and to clear away brush and flammable tinder. Fire season has arrived, and last year's drought has only increased the danger. You locate an external 4-plug electrical outlet on the wood facing of an exterior wall, and plug in your chainsaw, power drill, and a brush trimmer to find that they are all working properly. After installing the lights and waging a few long hours of brush warfare, you head back toward the house only to be hit with the hot odor of an electrical problem. You immediately check the one outlet you've been using, and discover all of the plugs are blistering hot, with the one connected to the brush trimmer partially melted. The adjacent exterior wall also feels warm to the touch. You head inside and feel that the corresponding interior drywall is even hotter and giving off whiffs of burning wood.

The Challenge: If the house bursts into flames, not only could you lose your own property, but given the conditions, it could easily develop into a full-blown forest fire that wipes out your neighbors and much of the wilderness area as well. Not exactly the low-impact, Thoreau-like experience you were looking for. So what do you do now?

Here's what you've got: Your hybrid SUV is well-gassed and ready to go, but your cellphone shows no reception here. In addition to the aforementioned tools and extension cords, the house and kitchen have the basic living essentials: furniture, pots and pans, etc. You also have a weekend's supply of food — but roasting it over your burning house is not really an option. And the wall is only getting hotter ...

Send a detailed description of your MakeShift solution with sketches and/or photos to makeshift@makezine.com by Nov. 30, 2007. If duplicate solutions are submitted, the winner will be determined by the quality of the explanation and presentation. The most plausible and most creative solutions will each win a MAKE sweatshirt. Think positive and include your shirt size and contact information with your solution. Good luck! For readers' solutions to previous MakeShift challenges, visit makezine.com/makeshift.

Lee D. Zlotoff is a writer/producer/director among whose numerous credits is creator of *MacGyver*. He is also president of Custom Image Concepts (customimageconcepts.com).

Photograph by Jen Siska

162 Make: Volume 11

Purely Platonic

A dodecahedron table lamp. By Charles Platt

People appear symmetrical, but even the most perfect human face shows irregularities if we compare the left side with the right. Perhaps this is why the absolute, rigid symmetry of crystals seems beautiful yet alien to us. Unlike DNA's soft spiral, a crystal's molecular bonds align themselves to form regular three-dimensional structures, which the Greeks considered mathematically pure. The most fundamental of these shapes are known as the five Platonic solids.

If you assemble equal-sided triangles — all the same size, with the same angles to each other — you can create three possible solids: a tetrahedron (with 4 faces), an octahedron (8 faces), and an icosahedron (20 faces). If you use squares instead of triangles, you can create only a hexahedron, commonly known as a cube. Pentagons create a dodecahedron (12 faces), and that's as far as we can go. No other solid objects can be built with all-identical, equal-sided, equal-angled polygons.

The Platonic solids have always fascinated me. My favorite is the dodecahedron, which is why I used it in this project as the basis for a table lamp. By extending its edges to form points, we make something that looks not only mathematically perfect, but perhaps a little magical.

START WITH A CRYSTAL

Because none of the Platonic solids, except for the cube, contains obvious 90° angles, building them is a counterintuitive, mind-bending experience. Before we get to the dodecahedron, let's warm up with something simpler: an octahedron.

You can make this in a few minutes using 12 plastic cocktail straws and 6 squares of duct tape, laying them out as in Figure B. Circle the straws around so that point A sticks to point B. The squares of tape should bend like hinges while the straws remain straight.

Now hinge the vertical straws so that their points C all meet together at point D. Again, keep the straws rigid, and flex the tape. Turn the structure upside down, bring points E to point F, and the result should look like Figure C. To prevent the straws from coming unstuck, you can bend the tape inward so that it sticks to itself.

Octahedrons are a common structure on the molecular scale, and because a crystal grows by repeating itself, tiny octahedrons assemble to form big ones. Search for "crystal octahedron" on eBay, and you'll discover that rockhounds know all about Platonic solids.

Notice how rigid your drinking-straw octahedron is. In fact, its shape is so efficient that it can support as much as 1,000 times its own weight. This suggests how rocks and metals achieve their strength.

FROM 8 TO 20 TO 12

Let's try something a little more permanent than drinking straws and duct tape. You'll need 8' of 10-gauge, solid copper wire, and 60 electrical ring terminals, size 12–10. (Within this terminal size specification, if you find a choice of ring sizes, select the ones with the smallest holes.)

Begin by hammering a couple of finishing nails, 3¾" apart, into a block of scrap wood. Now cut a piece of wire 3" in length, use pliers to pull the plastic shields off 2 ring terminals, and slip the terminals onto the ends of the wire. Place the assembly over the nails to hold everything in position, and solder the terminals onto the wire with a 30W (minimum) soldering iron.

After you do this 30 times, you'll have enough components to build the icosahedron shown in Figure D. You can use ⅜" #10 bolts to join the ring terminals, which you'll have to bend slightly to make them align with each other.

Now here's the interesting part: if you disassemble

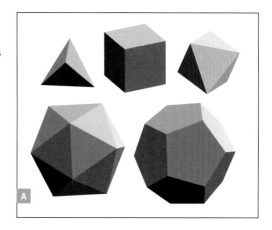

Fig. A: **The five Platonic solids: tetrahedron (4 sides), hexahedron (6 sides), octahedron (8 sides), icosahedron (20 sides), and dodecahedron (12 sides).**

the icosahedron, you can build the dodecahedron in Figure E using exactly the same number of pieces of wire because both solids have the same number of edges.

You'll find that the icosahedron is very easy to build, as the triangles cannot be deformed. The dodecahedron is very difficult because its pentagonal sides collapse easily. Therefore we'll need to fabricate our dodecahedral table lamp from a material that has its own rigidity, such as plywood or (my personal preference) ABS plastic (*see MAKE, Volume 10, page 100, "Plastic Fantastic Desk Set" for an intro to working with ABS*). Since we'll put a cool-burning fluorescent bulb inside the lamp to avoid overheating it, our dodecahedron must be big enough to contain the bulb. The suggested minimum dimensions are shown in Figure G.

If you're wondering how to cut a symmetrical pentagon, there's an easy way and there's a harder way. The easy way is to draw a pentagon using vector-graphics software such as Adobe Illustrator, then print the pentagon and use it as a template. (Old versions of Illustrator are cheaply available on eBay and will run on Windows versions up to XP.)

The harder way is to use a pencil, paper, and pro-tractor. Whichever way you do it, the inside angle at each point is 108°, and each side is angled 72° from each previous side. (*CRAFT magazine, Volume 04, page 96, "Repeating Splendor," has step-by-step instructions for drawing polygons with any number of sides.*)

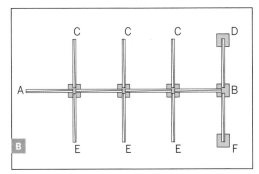

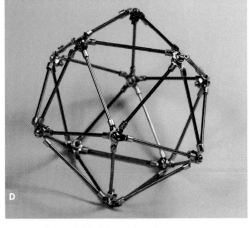

Fig. B: **An octahedron can be made from 12 plastic cocktail straws and 6 squares of duct tape.**

Fig. C: **Fold straws as shown to complete the shape.**

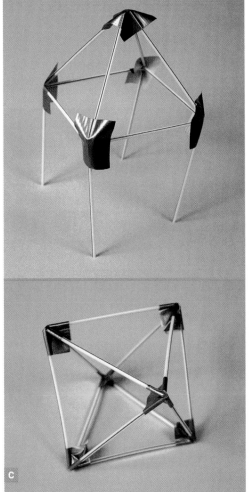

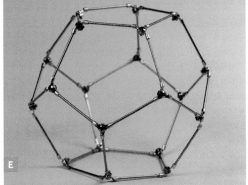

Fig. D: **Icosahedron made from 30 identical segments.**

Fig. E: **Dodecahedron made from the same segments.**

Fig. F: **The core of the lamp is made from ABS plastic.**

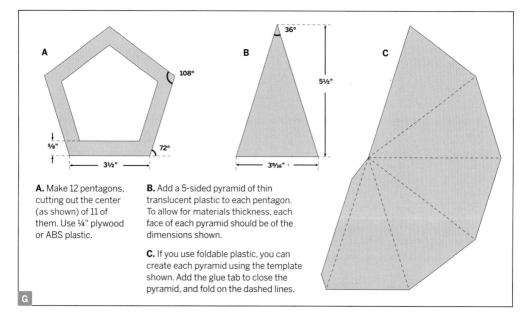

A. Make 12 pentagons, cutting out the center (as shown) of 11 of them. Use ¼" plywood or ABS plastic.

B. Add a 5-sided pyramid of thin translucent plastic to each pentagon. To allow for materials thickness, each face of each pyramid should be of the dimensions shown.

C. If you use foldable plastic, you can create each pyramid using the template shown. Add the glue tab to close the pyramid, and fold on the dashed lines.

CONSTRUCTING THE LAMP

You will need 12 pentagons. I used a band saw to cut them out of plastic sheet, but a handsaw is almost as quick. Bevel the edges with a sander or metal file so that when we add points to the solid (see below) they'll fit nicely. Use a jigsaw to cut out the centers of your pentagons, except for the last one, in which you will drill a hole for your lamp socket, before you mount it on the base of the lamp. I made the base by bending ABS, but you should use anything that appeals to you.

To assemble your pentagons, you can join them edge-to-edge using brackets made from aluminum sheet. My local hardware store had some very thin, precut aluminum, which was ideal. Use shears to take a few 1" strips from the sheet, chop each strip into pieces 2½" long, and bend each piece length-ways in a vise to make the brackets. A quick way to attach the brackets to the pentagons is with ⅛" aluminum rivets and a hand riveter tool.

ADDING POINTS

When your dodecahedron is complete, it should look like the one in Figure F. The final step is to stellate it, which means adding points to give it a star shape. Imagine the edges of each pentagon extended outward, and eventually they will intersect at points.

Each point will be a little 5-sided, hollow-based pyramid attached to one of the pentagons in the body of your lamp. You can make the pyramids

using 0.01" polycarbonate film with a velvet finish from mcmaster.com, vellum from your local craft store, or any other translucent paper or plastic that is foldable. Copy the 5-sided template in Figure G and fold along the lines to make one stellation. The only problem with this scheme is that the glue tab to complete the process will show as a shadow when the light is on. For a cleaner result you can fabricate each pyramid from five separate triangles using a thicker, rigid plastic (such as milky-white ⅛" acrylic) and join the triangles edge-to-edge with solvent glue or epoxy. Either way, to attach the stellations to the dodecahedron, use any clear adhesive such as 5-minute epoxy.

In the mathematical realm, the object you have constructed is considered a very pure and perfect form. In the physical universe, its perfection will depend on your proficiency with tools and glue. Either way, if people ask you what it is, you should have your answer ready: "Oh, that's just my stellated dodecahedron." And if someone needs to know what a dodecahedron is, you have another easy answer: "The fifth Platonic solid. What else?"

For more fun with Platonic solids, check out our Weekend Project on Picnic Geometry and learn how to make an icosahedron out of paper plates: makezine.com/go/picnic_geometry

Contributing Editor Charles Platt wrote "Plastic Fantastic Desk Set" for MAKE, Volume 10.

Friction. There's the rub.
By Donald E. Simanek

» Illusions and paradoxes never cease to fascinate.

When we perceive something that seems to behave in an unexpected or impossible way, we realize that it's something new to our experience and want to figure out what's going on. Visual illusions are the obvious example, but there are also tricks of physics and mechanics that seem paradoxical when first experienced. They challenge us to "puzzle out" what makes them work and are often the basis of magic tricks, toys, and illusions, or, at least, instructive physics demonstrations. We'll explore some of these tricks that may not be familiar to readers, with an emphasis on those that you can do or make yourself. I'll include explanations, as well as web links to more detailed treatments. Readers are invited to contribute their own favorites to me at dsimanek@lhup.edu.

Let us all give thanks for friction. Without it, our feet would slide uncontrollably when we try to walk, cars would spin their wheels and go nowhere, mountains would subside, and weather patterns would be vastly altered. Nearly everything in our world would function differently without friction.

Friction has the unfortunate side effect of dissipating kinetic energy, converting it into heat, which contributes to the inefficiency of machinery. Yet, without friction, it's hard to imagine how we could even manufacture machinery. Some friction is a good and useful thing. A lot of it is too much.

How many of us can claim we understand friction? The details of intermolecular forces and surface films are complex and won't be dealt with here. But in everyday life there are just a few basics you need to know.

A Little Friction Physics

When bodies are touching, they experience contact forces at their interface. These forces obey Newton's third law: If body A exerts a force on body B, then B exerts an equal-sized but oppositely directed force on A. The contact force acting on a body has two components: one perpendicular to the interface (called the "normal" force, symbol "N") and one tangential to the interface (called the "force due to friction," symbol "f"). If there's no sliding at the surfaces, the force due to friction can be anywhere from zero in size to a maximum value $f = \mu_s N$,

where μ_s is called the "static friction coefficient." If there is sliding, the force due to friction is given by $f = \mu_k N$, where μ_k is the "kinetic friction coefficient." Friction coefficients are nearly constant for a given interface. For most materials, μ_k is slightly smaller than μ_s. Both are usually less than 1, but for quite "sticky" surfaces, can exceed 1.

If there's no sliding of the bodies at their surfaces, the size and direction of the force acting on a body due to friction is just equal and opposite to the tangential component of the vector sum of all other forces acting on the body. If there is sliding, the force due to friction is in a direction opposite to the direction of the sliding (opposing the sliding motion). In short, the forces due to friction adjust in size and direction, responding to other forces so as to prevent sliding. But they have limits. When the limit is exceeded, sliding occurs, and the force due to friction acts in a direction opposing the sliding.

The Oscillating Beam Machine

This is an old physics homework problem. A heavy uniform bar or beam rests on top of two identical rollers that are continuously turning in opposite directions, as shown on the following pages. There's friction between the rollers and the bar, and the sliding friction coefficient is constant, independent of the relative speed of the surfaces. Find the motion of the bar. What happens if the rotation directions of both wheels are reversed?

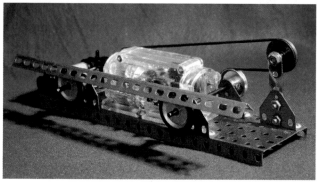

The front and back views, respectively, of the author's oscillating beam machine, made from standard Erector set components.

Many physics students have suffered through this problem, but few have bothered to make a working model of it. You can easily make it from construction set parts or whatever you have around the workshop. The rollers of my model are 1" pulleys, about 6" apart, with thick rubber O-ring tires. The rollers are driven by a pulley arrangement with a long rubber O-ring belt and a standard Erector set motor geared down to slow speed. The beam is a 10" or 12" angle girder.

Place the girder on the rollers so that it is an upside-down V. As the rollers turn, the girder oscillates back and forth in simple harmonic motion, without falling off the rollers. Sometimes the friction is a bit erratic, but the girder stubbornly refuses to fall off. I've had a model in a display case with a push button to activate the motor, and many people have tried to topple the girder without success. Of course, if the motor direction is reversed, both rollers rotate the opposite way, and the girder smoothly moves in one direction and falls off.

I used a motor from a 1970s Erector set powered by two 1.5V AA cells. This motor has a gear-reduction train built in. Modern sets have a motor with higher speeds and will require gears to reduce the roller speed to about 1 revolution/second. A rubber belt is best to transfer the motion to the rollers.

This is a fascinating toy to put on your desk and turn on whenever you wish to meditate on the futility of life. It could be a metaphor for some workplaces: lots of action, plenty of friction, but going nowhere and accomplishing nothing useful.

How does it work?

In the diagram, the normal forces are labeled N_1 and N_2, and the corresponding forces due to friction are f_1 and f_2. The sum of the two normal forces is always equal to the weight of the beam, W. The actual sizes of the normal forces depend on the position of the center of mass, and vary in size so that the sum of the torques on the beam is always zero.

If the center of mass of the beam were exactly

Photography by Donald E. Simanek

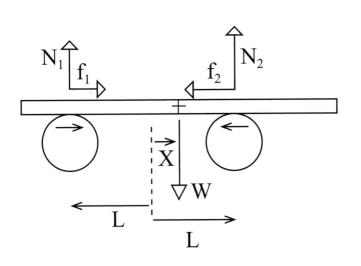

WHY DOESN'T THE BEAM FALL? The centers of the wheels are each a distance L from the centerline (dotted) of the apparatus. The friction forces are in opposite directions. When the beam is off center, one wheel supports a larger fraction of its weight, and the friction on that wheel is greater (f_2 in the example above). Because of the counter-rotation of the wheels, the net friction force is always toward the central position. This slows the beam as it moves away from center, then forces it back toward center. The beam overshoots center, and the action repeats on the other side. **QUESTION: What happens if both wheels turn the other direction?** See **makezine.com/go/friction** for the answer.

between the rollers, the load on each roller would be the same. If the friction coefficients were the same at each roller, the force due to friction would be the same at each roller. Then, the beam would not move, and the rollers would slip under the beam. But such perfection is never attained. The initial position of the beam is not exactly centered, and the friction at the rollers is not uniform.

Suppose the beam begins to move toward the right roller. The right roller then takes a greater fraction of the load (the weight of the beam), the friction at the right roller increases in proportion, and the force due to friction opposes the beam's move to the right. If the beam moves to the left and overshoots the central position, the left roller takes more of the load, and the left roller's friction forces the beam back toward the right. Whichever direction the beam moves, due to any unbalance, the rollers act to force it back toward the center.

But the motion toward the center invariably causes it to "overshoot" the central position, and the back and forth motion continues. This example has something in common with a pendulum, as a more mathematical analysis would show.

You could try to center the beam precisely and start it so as to achieve no oscillation of the beam. Good luck! Let me know when you succeed. There's a physics mystery here that you won't discover until you build and play with this machine. Why does the slightest off-center initial position build up to a situation where there's a larger amplitude of beam oscillation than the initial displacement? What determines that amplitude? And why doesn't the beam ever come to rest? We leave these questions as exercises for the reader.

For those interested in the mathematical details, see: makezine.com/go/friction.

Donald Simanek is emeritus professor of physics at Lock Haven University of Pennsylvania. He writes of science, pseudoscience, and humor at www.lhup.edu/~dsimanek.

Here's my adaptation of the

ORANGE CRATE RACER PROJECT

FROM THE 1949 CLASSIC *MAKE IT AND RIDE IT* BY C.J. MAGINLEY, ILLUSTRATED BY ELISABETH D. McKEE

There were endless jewel thefts, art forgeries, safe crackings, land swindles, and train robberies in my suburban childhood town, but my lack of transportation and an office kept me from cracking a single case. I was certain that if I had a trusty '40 Ford, the right overcoat, and a defendable treehouse, I would be able to run a successful agency. Despite having Private Investigator business cards, clients just wouldn't take me seriously without the appropriate ride. Though I would not have been much closer to my noir dreams, it would have been nice to obsess over *Make It and Ride It* with my Pinkerton-certified magnifying glass.   —Mister Jalopy

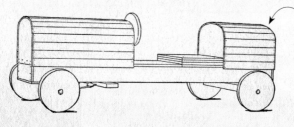

If the trunk were left open, it would be an ideal seat for a terrier named Dr. Watson.

MATERIALS:

2 pc two by four 62" long — frame
2 pc ¾" × 62" × 11½" — end pieces X, Y
1 pc ¾" × 1¾" × 20" — footrest
15 orange crate ends or partitions — floor, bulk-heads, and frame pieces A, B ◄——
4 pc two by two 11½" long — reinforcing pieces
2 pc two by four 16" long — axletrees
4 wheels 10" in diameter

One ¾" disc 8" in diameter — wheel
One ¾" disc 3" in diameter
1 pc ¾" dowel 18" long — shaft
Two 4" eye bolts
5'-6' ⅛" wire cable
Two ¾" awning pulleys
2 pc ¼" dowel — pegs

10 flat iron washers about 1" in diameter
Two flat-headed screws 1½" long
One ⅜" bolt 4"-5" long — king bolt

Though not listed in the Materials section, the text suggests using orange crate side slats, linoleum, canvas, or sheet metal for the body.

No orange crates in your attic? Crate ends measure 10½"×11½" and ⅝" thick.

STEERING ASSEMBLY

1. FRAME

BOTTOM VIEW

front axletree

FRONT VIEW

front axletree

2. AXLETREES AND FOOTREST

king bolt

TOP VIEW

rear axletree

footrest

front axletree

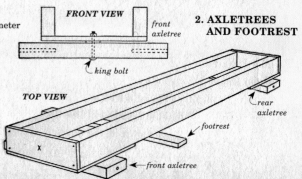

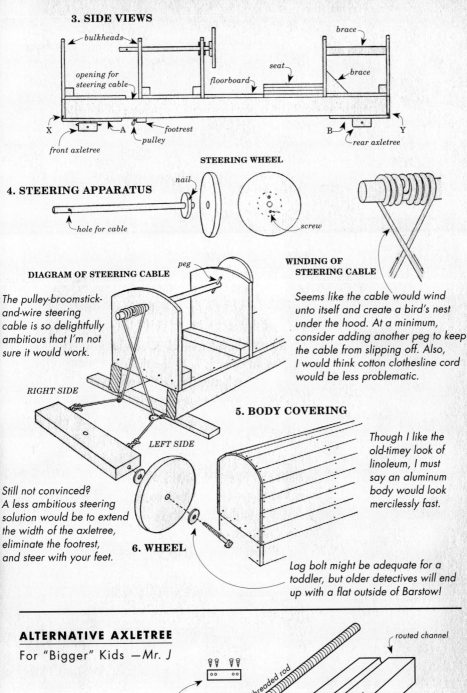

3. SIDE VIEWS

bulkheads

brace

brace

opening for steering cable

seat

floorboard

X   A   footrest   B   Y

pulley

front axletree   rear axletree

STEERING WHEEL

nail

4. STEERING APPARATUS

hole for cable

screw

WINDING OF STEERING CABLE

DIAGRAM OF STEERING CABLE

peg

The pulley-broomstick-and-wire steering cable is so delightfully ambitious that I'm not sure it would work.

Seems like the cable would wind unto itself and create a bird's nest under the hood. At a minimum, consider adding another peg to keep the cable from slipping off. Also, I would think cotton clothesline cord would be less problematic.

RIGHT SIDE

LEFT SIDE

5. BODY COVERING

Though I like the old-timey look of linoleum, I must say an aluminum body would look mercilessly fast.

Still not convinced? A less ambitious steering solution would be to extend the width of the axletree, eliminate the footrest, and steer with your feet.

6. WHEEL

Lag bolt might be adequate for a toddler, but older detectives will end up with a flat outside of Barstow!

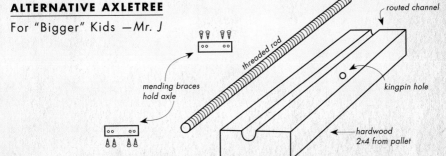

ALTERNATIVE AXLETREE
For "Bigger" Kids —Mr. J

routed channel

threaded rod

mending braces hold axle

kingpin hole

hardwood 2×4 from pallet

Skip to the beat of a different drummer, eliminate red-eye the old-school way, and learn about the origins of invention.

TOOLBOX

Hop on Water
AquaSkipper
$500 inventist.com

Here's a new personal watercraft that doesn't burn gasoline, pollute the water, or make noise. It's a human-powered hydrofoil ornithopter. Got that?

It's a spidery aluminum contraption with wings that fly through the water as you jump up and down on the platform above. You jump off a dock with it and skim over the top of the water.

If you're good enough, you ride it back to the dock and jump off again. If you're not that good yet, you wipe out and swim it back to the dock for another try. It's great exercise and lots of fun.

—*Tim Anderson*

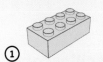

①

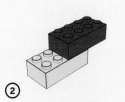

②

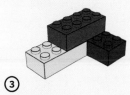

③

Never-Ending Bricks
Lego CAD with LDraw
Free ldraw.org

Quick, grab three 2×4 Lego bricks. Place the second brick on top of the first brick, offset by two studs on the long axis. Now, place the third brick under the second brick, at a right angle to the first brick. Huh?

This kind of tortured description made me wonder about making my own visual Lego instructions. I Googled my way to the free LDraw suite of Lego CAD applications and parts libraries, which are staggeringly complete. Soon after installation, I was using the MLCad application for a marathon 3D brick-building session. It's addictive to drag-and-drop any Lego part ever made from an endless tub of virtual bricks.

Pass the step-inclusive model file to the free LPub program, and you'll soon be printing your own Lego instruction manual and parts list. Thanks to the bundled POV-Ray renderer, you can choose from simple, flat graphics to shiny, ray-traced works of art. —*John Edgar Park*

Go Juice
Pumice Hand Cleaner
$20 for 1 gallon gojo.com

Everyone has their favorite cleaner, and Gojo is ours. We like the mild orange scent, the abrasive action of the pumice, and the fact that it doesn't leave your hands feeling greasy or dry.

We keep a small bottle at one sink, which we refill from the gallon pump bottle at the other sink. One excellent design feature is that each bottle comes with a handy nailbrush, which clips onto the side of the bottle.

Gojo is great after you've packed your bearings and you're ready to get the grease off your hands. It really shines for cleaning up after fixing a flat on your bicycle. It can be used for cleaning dirty and greasy parts as well. —*Lenore Edman*

Carry That Weight
Bike Panniers
$130+ ortliebusa.com

A bunch of my coworkers and I bike to the San Francisco-Alameda ferry almost every day, and several of us use Ortlieb panniers. They are built to be waterproof (this is the same company that makes the dry bags I've used on sea kayak trips), are easily taken on and off the bike, and are convenient for carrying everything (cartons of milk, clothes, spare tubes, laptop, bike lock ... all at the same time). One accessory I would add to my current setup is the little bag called the In-Put, which would help me locate my keys, phone, and wallet rather than having to dig through all the detritus of my life to find the essentials.

—*Andrea Dunlap*

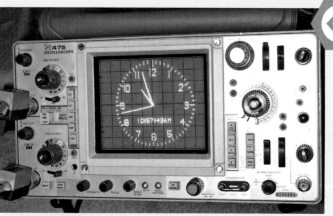

Oscilloscope Clock Kit

$35 dutchtronix.com/ ScopeClock.htm

This is a fantastic-looking kit, turning an oscilloscope into a clock! It uses all through-hole components, but you need to have your own oscilloscope. (Who doesn't, really?)

Outdoor Kits

Prices vary thru-hiker.com/kits.html
Thru-Hiker has some nifty kits and DIY projects for folks who like to not only get out-of-doors, but make some of their own gear — everything from your own sleeping bag to the sack you stuff it in, a windproof shell, or a down vest.

MPA: MIDI Decoder Kit

$48 highlyliquid.com/kits/mpa
This MIDI Decoder from Highly Liquid is for DIY filter circuits, circuit bending, and MIDI-to-DIN sync conversion.

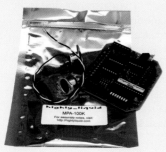

Tube Amp Kit

Free diyaudioprojects.com

Featuring several DIY audio projects for the electronics hobbyist and woodworker, this site has a range of tube amp kit instructions and some interesting mods, as well as good discussions.

« Brooks Leather Saddles www.brookssaddles.com

These leather saddles conform to your body and your riding style as you ride. They do take some breaking in, so you can't judge them on a test ride. Keep the leather supple and they will last forever. When you see those 50-year old bikes with the ancient saddles still in use, those are Brooks!

« Xtracycle xtracycle.com

This great attachment for pretty much any bike extends the back end by a couple feet, and adds two large side slings for carrying gear and a large rear platform. It's lightweight and quite stable, considering it can carry close to 200lbs, and is a great alternative to a trailer. *Momentum* graphic designer Chris Benzen has his on an old mountain bike frame set up for road riding with fenders and 26×1.25 tires. He's carried everything from camping gear to another rider with a flat tire!

« Knickers swrvecycling.com, chromebags.com

Also known as "shpants," these mid-calf-length pants stay clear of the bike's greasy drivetrain. The latest style trend to borrow from: messenger chic. There are many new brands and styles to choose from, including some made from stretchy and/or water-resistant fabrics.

Bungee Cords

These elastic wonders with hooks at both ends truly are a cyclist's best friend. Whether we're securing a box onto our rack or creating a gigantic sculptural assemblage in our trailer, bungee cords get the job done. The beauty of bungees is that they come in a variety of lengths and thicknesses (and colors!), and there are a million ways to combine and tie them. Bungee tying is an art form!

« Dyno/Generator Hub sram.com

Battery-powered, clip-on lights are easily stolen, their mounting hardware usually breaks, and they consume a lot of batteries. With a Sram i-Light hub, you never need to worry about theft or dead batteries. Yes, it's relatively expensive and heavy, but after three months, Ecstatic Mechanic columnist Omar Bhimji is still taking dark routes home and aiming at walls just to revel in the joy of lighting his own way.

Honorable mentions go to a variety of useful items:

Zap straps, steel bicycle frames, bike computers (yay, mileage!), rain booties, clipless pedals, and ... cornstarch, for lubricating inner tubes if you don't have talcum powder handy, plus it's good for making crunching snow sound effects for your indy movie or radio play, not to mention cooking.

Momentum provides self-propelled people with inspiration, information, and resources to help them fully enjoy their urban biking experience and connect with their local and global cycling communities. momentumplanet.ca

« Web Master

Rule the Web by Mark Frauenfelder
$15, St. Martin's Griffin

Finally, a computer book for the masses. Clear, concise, and to the point, Frauenfelder uses easy language to casually explain absolutely everything you'd ever want or need to know about utilizing the internet more effectively. At first glance I was like, "Great, another unusable computer guide book." After picking it up, though, I went from randomly flipping through pages to actually being pleased at the idea of reading it cover to cover.

Step-by-step practical applications show how to deal with everyday annoyances such as those endless pre-approved credit card offers. He even includes eye-opening information on the security loopholes in the issuance of these new cards and how to protect yourself against them. —*Walter Kalwies*

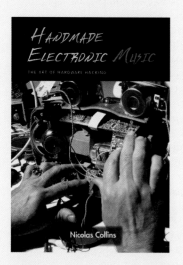

« The Sound and the Fury

Handmade Electronic Music by Nicolas Collins
$26, Routledge

Here we have, at last, an electronics book that caters to people who have ideas first, and electronics second. A former technician to some of the last century's most imaginative experimental composers, Collins offers a splendidly integrative look into the history of "sound art," basic electronics, and junk re-visioning.

The book teems with deliciously messy sound projects adapted for battery power (banishing electrocution worries, and allowing for proper experimental mischief), from how to turn your wall into a speaker, to how to extract sounds from remote controls. Collins will have you looking to your previously mundane surrounding electronic elements for uses only you can dream up. —*Meara O'Reilly*

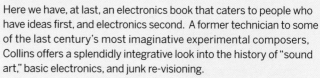

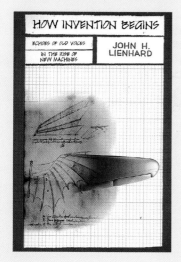

« Invent This

How Invention Begins by John H. Lienhard
$30, Oxford University Press

This is a very readable account of how inventions arise — quietly, without "aha!" moments, and seldom with clear priority for one inventor. Flying machines, the quest for speed, and the early history of mechanical printing are just a few of the topics Lienhard examines. Often the real value of an invention is different from the motivation that produced it. The printing press eliminated much of the hand labor previously required for book production, but the real (and perhaps unintended) outcome was mass education and a proliferation of schools and libraries. Lienhard writes in an engaging style, delighting in telling a good story while still keeping it honest. —*Donald Simanek*

10 Best Books on Human-Powered Vehicles

These books collectively give the best intuition for energy consumption you'll ever get, and they illustrate the relevance of mechanical efficiency and good design with the tightest constraints. Many of my heroes are represented somewhere in this list.

—*Saul Griffith*

Human-Powered Vehicles
by Allan V. Abbott and David Gordon Wilson

$200, Human Kinetics Publishers

David Gordon Wilson is my hero. This is the definitive text on human-powered vehicles.

The Bicycle Wheel
by Jobst Brandt

$25, Avocet Publishing

This book is simply sublime. It has all the fundamentals of wheel building, and the physics and science are well-described.

Gossamer Odyssey: The Triumph of Human-Powered Flight
by Morton Grosser

$20, Zenith Press

A great history of technical triumph in the hardest of all human-powered disciplines: flight.

Bicycling Science, 3rd Edition
by David Gordon Wilson

$25, MIT Press

It's been reviewed here before, but there is simply no better book on bicycles and the biomechanics of human power.

The Strip-Built Sea Kayak: Three Rugged, Beautiful Boats You Can Build
by Nick Schade

$20, Ragged Mountain Press

I love the elegance and simplicity of these designs. The kayak deserves such a fascinating book.

Birdflight as the Basis for Aviation
by Otto Lilienthal

$20, Markowski International

This is where it all started. Lilienthal died doing what he loved, furthering human flight.

Atomic Zombie's Bicycle Builder's Bonanza
by Brad Graham and Kathy McGowan

$25, McGraw-Hill

Nothing elegant here — it's all hacks, just raw, good fun to be had with a welder. Every maker would be inspired by it.

Classic Small Craft You Can Build
by John Gardner

$25, Mystic Seaport Museum

For the craftsman with appreciation for human-powered wooden boats.

Bike Cult: The Ultimate Guide to Human-Powered Vehicles
by David B. Perry

$40+, Four Walls Eight Windows

More of a coffee table book for history and inspiration than reference, it's great for browsing.

The Immortal Class: Bike Messengers and the Cult of Human Power
by Travis Culley

$6, Random House

If you've ever ridden a fixie or a single speed, you will love this book. If you've ever cursed a messenger, you should read this for empathy.

A Slave 4 U

$35 bhphoto.com

Today's whiz-bang point-n-shoots do everything but fold your clothes. But they're a giant step backward in one way: flash photos. A tiny flashtube one inch from your lens guarantees demonic red eye and a harsh, blasting light with all the charm of a driver's-license photo. Any pro shooter will tell you, you need light from the side to bring out shapes and textures. But how to do that without a pack-mule's worth of equipment?

Set Vivitar's tiny DF120 digital slave flash to the side of your subject, where its photocell can see your camera's built-in flash. Unlike most optical slaves, this one has extra settings to ignore the rapid preflashes many digital cameras use; it triggers only for the actual photo. My tip: Drape some tissue or milky plastic over your camera's flash, then try dialing in -1 or -2 exposure compensation: your shirt-pocket portrait studio is ready to go. Beautiful, dahling, give me more! —Ross Orr

Retro-Robot Art!

Prices vary ericjoyner.com

Those of you old enough to remember the tin robots of the 50s and 60s or the Rock 'Em Sock 'Em Robots of the early 70s, or any robot aficionado for that matter, will appreciate Eric Joyner's awesome take on these classic toys in his highly original paintings.

Check out an animation of his paintings on makezine.com/go/joyner, then visit his store to get these cool prints on everything from tees to coffee mugs. What's with the donut and robot theme? "Two of my favorite things," he replies. Mine too! Mmmm, donuts ... —Rob Bullington

National Ornamental Metal Museum

metalmuseum.org

I strongly recommend the metal-working classes at the National Ornamental Metal Museum in Memphis, Tenn. Virtually every major artist-blacksmith in America has trained or exhibited here (seriously, just ask them). The museum offers both beginning and advanced courses in blacksmithing, copper wind vanes, metal casting, and many other subjects.

The museum is heavily involved with the current artist-blacksmith movement, and is (I believe) the only museum in the U.S. devoted entirely to ornamental metalwork. The director, Jim Wallace, started this place from scratch after college. With its stunning location on the banks of the Mississippi, I believe this is one of the undiscovered treasures of the United States. —Darel Snodgrass

Andrea Dunlap is a filmmaker and photographer for seedlingproject.org.

Darel Snodgrass is a professional classical radio announcer and an amateur metalworker.

Donald Simanek's "Museum of Unworkable Devices" can be found at www.lhup.edu/~dsimanek.

John Edgar Park (jp@jpixl.net) is a character mechanic at Walt Disney Animation Studios.

While **Lenore Edman** is neither evil nor mad, she is one of the Evil Mad Scientists at evilmadscientist.com.

Meara O'Reilly (myspace.com/mearabai) detunes pianos and tinkers with transducers.

Ross Orr keeps the analog alive in Ann Arbor, Mich.

Saul Griffith has a not-so-secret love affair with the bicycle.

Tim Anderson has a home at mit.edu/people/robot.

Walter Kalwies is an interstellar rubberneck ready to break out into spontaneous revelry.

Have you used something worth keeping in your toolbox? Let us know at toolbox@makezine.com.

MAKE's favorite puzzles. (When you're ready to check your answers, visit makezine.com/11/aha.)

Point of Gnome Return

One hundred very smart garden gnomes are snatched from their homes by an evil wizard. He tells them he is going to line them all up in a row, and place a red or blue hat on each of their heads. They won't be able to see the color of their own hats or anyone's behind them, but they will be able to see the hats of the gnomes in front of them. The wizard will start at the back of the line and ask each gnome to guess the color of his own hat. Each gnome can only answer either "red" or "blue." If he gives the wrong answer, he will be led off to work on the wizard's landscaping for the rest of eternity. If he answers correctly, he will be returned to his own garden. Then the wizard will move on to the next gnome in line.

All of the gnomes will be able to hear the answers of the gnomes behind them, but they will not know if they were led off to forced labor or if they answered correctly and were set free. The gnomes are allowed to consult and agree on a strategy beforehand (while the wizard listens in) but after being lined up, they will not be able to communicate in any other way besides their answer of "red" or "blue" (in other words, they won't be able to change the pitch of their voice or give any other clues to the other gnomes once they are in line and the hats are placed on their heads).

What strategy should the gnomes use to maximize the total number of gnomes that will be set free?
Hint: They can do pretty well, even if the wizard hears their plan and puts the hats on in such a way to thwart whatever idea they come up with.

Human Race

Five makers (Ben, Mark, Jason, Ruth, and Joel) raced in an extravaganza for alternative transportation. They competed in five separate races: a Segway race, a bicycle race, a hybrid car race, a skateboard race, and a roller skate race. Points were given as follows: 5 points for first place, 4 points for second, 3 points for third, 2 points for fourth, and 1 point for last place. Overall scores were determined by adding up the individual scores from each race.

Ben got the highest number of points: 24. Jason finished in the same place in four out of the five races. Joel came in first in the skateboard race, and third in the roller skate race.

The final total rankings for the races were: Ben, Mark, Jason, Ruth, and then Joel. There were no ties in any individual races, and no two racers had the same total score.

What place did Mark come in during the skateboard race?

Michael Pryor is the co-founder and president of Fog Creek Software. He runs a technical interview site at techinterview.org.

The 1977 Gulf of Alaska Baidarka Expedition
By George Dyson

Thirty years ago, the now-ubiquitous sea kayak was a rarity on the Pacific Northwest coast. The fleets of Aleut baidarkas that had swept through Southeast Alaska in the 19th century had vanished without a trace. In 1977, a small group of boatbuilders decided to bring the baidarka back to life ...

1. Two hundred years ago, the coastline of Alaska was governed by the Russian-American Company, a monopoly chartered in 1799 by Emperor Paul under the posthumous instructions of his mother, Catherine the Great. The Russian colonization of North America, beginning with Vitus Bering's and Alexei Chirikov's explorations in 1741, and driven by the trade in sea otter furs with China, was characterized by the adoption (and adaptation) of indigenous technology, especially the skin boat. Under Russian supervision, fleets of as many as 700 one-, two-, and three-hatch kayaks (termed *baidarkas* by the Russians) made annual voyages of several thousand miles from their home bases in the Aleutian archipelago and on Kodiak Island to hunt sea otters along the Southeast Alaskan coast.

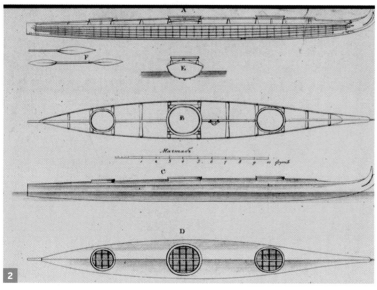

2. The three-hatch baidarka, ranging from about 22 to 30 feet in length, served as the light utility vehicle of the Russian American colonies, carrying non-paddling passengers, as well as equipment ranging from small cannons to medical supplies to the ubiquitous "portable church" — a sturdy, canvas tent that was fully equipped for Russian Orthodox services in remote settlements and even included roll-up icons that could be stowed within the boat.

Lithograph by F.A. Pettit, 1906, courtesy George Dyson; engraving from Langsdorff's *Voyage Around the World* (1812) after a drawing by Ivan Koriukin, ca. 1803, courtesy Bancroft Library

3. In early 1977, your author, along with friends Peter Johnston, Lou Kelly, Joe Ziner, and Harry Williams, began building a fleet of six 28' three-hatch baidarkas (similar to the 25-footer pictured here) for a voyage retracing the explorations of the Russians on the Gulf of Alaska coast. We used ½" OD 6061-T6 aluminum tubing for the transverse ribs and ¾" tubing for the longitudinals. The finished skeletons weighed about 35 pounds, and even at today's retail prices, it would only take about $200 worth of aluminum tubing to build the boat.

4. The "flatware" forming the bow and stern assemblies was made from 6061-T6 aluminum sheet salvaged from road signs purchased from North Star Salvage in Vancouver, B.C. We had enough signs left over to make paddles and rudders, too. We were able to make two paddle blades and two rudder blades from one stop sign, as pictured here.

5. After a voyage north from Vancouver, B.C., aboard the retired halibut schooner *Betty L.*, we launched the six baidarkas into the Gulf of Alaska on the outer coast of Chichagof Island on May 3, 1977, near the entrance to Lisianski Strait. This is where the first landing party under Alexei Chirikov had disappeared upon going ashore in search of water on July 15, 1741.

6. On a glass-calm morning in the middle of Chatham Strait, between Admiralty and Baranof Islands, Joe Ziner claims he saw something very large swim up to the surface and stare at him. The water here is 440 fathoms (2,640 feet) deep. Joe is staring back.

Photography and illustration by G. Dyson

7. Small square sails were sometimes used on two-hatch and three-hatch baidarkas in the 19th century. For covering long distances with minimal effort, sails are hard to beat. Our semicircular sails were only for sailing downwind. We found that we could only sail about one quarter of the time, but that about half the miles made good were under sail.

8. Near the mouth of Peril Straits, Baranof Island, we took shelter under our sails. The baidarka is an amphibious vehicle. We led a nomadic, amphibious life.

9. In the Gulf of Alaska, with a light westerly off Cape Spencer, we made about six knots under sail. In stronger winds, our baidarkas would surf downwind at speeds of 12 knots or more. A Japanese observer, after seeing Aleut/Russian baidarkas off Japan's Northern Islands in the 19th century, wrote, "The boat seems to fly over the waves as the bird flies in the sky."

10. Author George Dyson is a historian of technology who is trying to complete a book about the dawn of the digital universe and return to building boats.

Photography by G. Dyson (top, middle) and Thomas Macy (bottom); illustration by G. Dyson, sketch inspired by Joe Ziner

BARE NECESSITIES

FROM MAKER TO MAKER

Who doesn't appreciate a really good tip now and then? Especially the kind, as one reader put it, "that changes your life." Whether it's as practical as using a cooler as a light tent, or something more creative like making breakfast with your espresso machine, we all rely on our friends and neighbors to tip us off to the new and the good. —*Arwen O'Reilly*

Eggs-presso

MAKE columnist Saul Griffith extols the virtues of his local café: "I was floored when I noticed them scramble their eggs with the milk steamer of their espresso machine. I tried it at home, and it worked perfectly — my world is changed."

Let There Be Light

Check out makezine.com/go/cooler for the Strobist post about this no-fuss light "tent" that anyone can put their hands on. If you've got a cooler, then you have the perfect place to shoot seamless backgrounds to highlight your latest creation.

Rent Free(dom)

Brady Forrest raves about rentometer.com: "This is the type of site that shows why map mashups are so popular; when the right data is displayed on a map, it just comes to life. Rentometer will let you know if you're paying too much, too little, or just the right amount of rent for your unit type and area. Its incredibly clear and straightforward interface makes it a breeze to use."

And introducing a new, regularly appearing visual how-to ...

Tricks of the Trade By Tim Lillis

Don't ever lose a beer to gravity again!

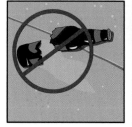

Use this trick from the corner store guy to keep your 6-pack from poking through a plastic bag.

Firmly grip the 6-pack and squash a corner on the counter, floor, table, or other hard, flat surface.

Squash all 4 corners and place in bag.

Show up at the BBQ with your pride and all your beers intact!

Have a tip for MAKE readers? Send it to tips@makezine.com. Have a trick? Submit to tricks@makezine.com. Also be sure to check out TNT, our Tools-N-Tips newsletter, at makezine.com/tnt.

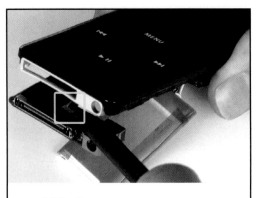

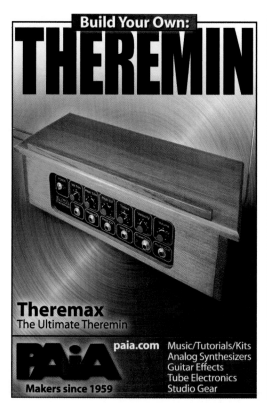

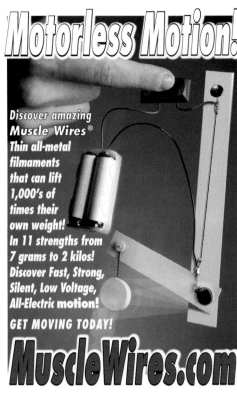

MAKER'S CALENDAR

Compiled by William Gurstelle

Our favorite events from around the world.

Jan	Feb	Mar
Apr	May	Jun
July	Aug	Sept
Oct	Nov	Dec

» SEPTEMBER

»Oceana Airshow
September 7–9
Virginia Beach, Va.
One of the world's largest air shows, it will include aircraft such as the F/A-18C Hornet, the E-2C Hawkeye, the F-15 Eagle, the F-22 Raptor, and a large-formation fleet flyby, as well as the U.S. Navy's Blue Angels precision flying group.
oceanaairshow.com

»British Association for the Advancement of Science Festival
September 9–15, York, U.K.
From lectures to hands-on activities to debates, this science and technology festival offers something to interest just about any type of scientific mind.
makezine.com/go/baasf

⚔ Reno Air Races
September 12–16, Reno, Nev.
Five days of racing by many different types of high-speed aircraft, plus an air show featuring the Canadian Forces Snowbirds.
airrace.org

»Sunfest Kite Festival
September 20–23
Ocean City, Md.
Kiting enthusiasts and spectators from across the country meet for four days of kite competitions, games, mass ascensions, and maker workshops. makezine.com/go/sunfest

»Robothon
September 21–23, Seattle
Robothon showcases some of the best of the amateur robotics community. People from around the world come to meet other enthusiasts, show off their robotic creations, and join in competitions.
robothon.com

»Renewable Energy Festival
September 22–23
Kempton, Pa.
This green living festival features exhibits, classes on making clean energy, workshops, and more.
paenergyfest.com

»Burning Man
August 27–September 3
Black Rock Desert, Nev.
A gigantic, week-long counter-cultural festival featuring pyrotechnical art, odd machines, and lots of interestingly repurposed technology.
burningman.com

Red Bull Soapbox Race

September 29, Seattle

This non-motorized racing event invites auto amateurs, gearheads, and adrenaline junkies to craft outrageous human-powered carts to compete in a downhill sprint against the clock. redbullsoapbox usa.com

Photography by Martin Trenkler/Red Bull Photofiles

» OCTOBER

Trinity Site

October 6

Near Alamagordo, N.M.

White Sands Missile Range offers a rare opportunity for the general public to visit Trinity Site, the birthplace of the atomic bomb. www.wsmr.army. mil/pao/TrinitySite/ trndir.htm

Space Shuttle Launch

October 20

Kennedy Space Center, Fla.

Space shuttle *Discovery* blasts off on the 23rd U.S. mission to the International Space Station. kennedyspace center.com/launches

+ See makezine.com/11/readerinput for letters from our readers offering praise, brickbats, and swell ideas. We'll Make Amends there, too.

Maker Faire

October 20–21

Austin, Texas

MAKE and CRAFT magazines bring Maker Faire to Austin. Maker Faire combines science, art, craft, and engineering into an exciting two-day extravaganza of making. makerfaire.com

Wirefly X-Prize Cup

October 27–28

Las Cruces, N.M.

The X Prize Foundation and Holloman Air Force Base host what they bill as "The World's Largest Air and Space Flight Demonstration." The event features multiple Lunar Lander rocket flights and a big-time air show. xprizecup.com

» NOVEMBER

World Championship Punkin Chunkin

November 2–4

Bridgeville, Del.

Hundreds of catapults, air cannons, and other hurling machines gather in deepest, darkest Delaware to determine who can make a machine that can hurl a 10-pound pumpkin the greatest distance. punkinchunkin.com

» Related story on page 30

DARPA Grand Challenge

November 3, Location TBA

Entrants compete in the DARPA Urban Challenge, featuring autonomous ground vehicles conducting military operations in a mock urban area. First place takes home a whopping $2M! darpa. mil/grandchallenge/ spectators.asp

WorldSkills Competition

November 14–21

Shizuoka, Japan

Nearly a thousand young people drawn from 45 countries will show off their skills in the fields of computer-aided manufacturing, computer-aided design, welding, mechatronics, polymechanics, and more. worldskills.org

IMPORTANT: All times, dates, locations, and events are subject to change. Verify all information before making plans to attend.

Know an event that should be included? Send it to events@ makezine.com. Sorry, it is not possible to list all submitted events in the magazine, but they will be listed online.

If you attend one of these events, please tell us about it at forums.makezine.com.

My Train-Schedule Alarm Clock

By Greg McCarroll

■ **It was a cold winter morning and I was** sitting on an even colder bench on a railway platform in London, cursing myself for hitting snooze. Because of that moment of weakness, I had to wait around for the next train. Did I mention it was cold?

So I started thinking about solutions — after all, I had time to kill. A 30-minute snooze function? No good, the gap between trains wasn't always 30 minutes. Any alarm clock would have to be aware of the railway timetable, or better still, be aware of delays.

That night, I thought about what I'd need to build such an alarm clock. I could figure out how to do almost everything: the snooze button, the music to wake me up, the web lookups — but I couldn't figure out how to easily build a display for the current time.

A couple of weeks passed until I ended up at a dinner party talking about the idea. A friend, Kate Pugh, pointed out the obvious — I already had an alarm clock with a display, so why not just use it?

The very next day, I built a new alarm clock that has a snooze button made mainly from Legos and an old mouse (my favorite bit was the wonderful contact that a thumbtack, pushed upside down into the Blu Tack-filled base of a Lego brick, made with the microswitch of the mouse). The snooze connects by a long cable to my Linux workstation, my old alarm clock, an AirPort Express, and a program quickly hacked together in Perl.

The program was made a lot easier by using various bits of code from CPAN (cpan.org) where possible. (I even managed to give a little back by releasing the module that I wrote to screen-scrape the train information.) To be honest, it was just a simple state machine that slept until it got close to wake-up time, then started monitoring the departure boards.

So I now have an alarm clock that wakes me a little later if trains are delayed, and works out the length of time to let me snooze based on the next departure needed to get me to work. Best of all, my employer agreed that if all the trains were canceled, the clock would email my workplace that I'd be working at home, and let me sleep in.

Greg McCarroll is a Perl hacker living in London with his wife, Ron, and cat, Hobbes. He loves technology that makes people's lives better, even if better means more time sleeping.

Photograph by Greg McCarroll